Synthesis of Digital Designs
from Recursion Equations

ACM Distinguished Dissertations

1982

Abstraction Mechanisms and Language Design
by Paul N. Hilfinger

Formal Specification of Interactive Graphics Programming Languages
by William R. Mallgren

Algorithmic Program Debugging
by Ehud Y. Shapiro

1983

The Measurement of Visual Motion
by Ellen Catherine Hildreth

Synthesis of Digital Designs from Recursion Equations
by Steven D. Johnson

Synthesis of Digital Designs
from Recursion Equations

Steven D. Johnson

The MIT Press
Cambridge, Massachusetts
London, England

This book was printed and bound in the United States of America

PUBLISHER'S NOTE: This format is intended to reduce the cost of publishing certain works in book form and to shorten the gap between editorial prepartion and final publication. Detailed editing and composition have been avoided by photographing the text of this book directly from the author's text-processor output.

Dissertation submitted to the faculty of the Graduate School in partial fulfillment of the requirements for the degree of Doctor of Philosophy in the department of Computer Science, Indiana University, May 1983. The thesis work was supported in part by the National Science Foundation, grant number MCS77-22325.

Library of Congress Cataloging in Publication Data

Johnson, Steven Dexter.
    Synthesis of digital designs from recursion equations.

    (ACM distinguished dissertations)
    Thesis (Ph.D.)—Indiana University, 1983.
    Bibliography: p.
    Includes Index
    1. Digital Electronics—Data processing. 2. Electronic circuit design—Data processing. 3. Recursive functions.
I. Title II. Series
TK7868.D5J64 1984   621.3815'3   83-25632
ISBN 0-262-10029-0

Series Foreword

This book is being published by The MIT Press as an outgrowth of the annual contest for the best doctoral dissertation in computer-related science and engineering. The contest was initiated in 1982 and is co-sponsored by ACM and The MIT Press.

The Distinguished Doctoral Dissertation Series has been created to recognize that some of the theses considered in the final round of selecting a contest winner also deserve publication. In the judgment of the ACM selection committee and The MIT Press, this thesis is of such high quality that it deserves special recognition in this new series.

Dr. Steven Johnson wrote his thesis, "Synthesis of Digital Designs from Recursion Equations," at Indiana University, and the thesis was submitted to the 1983 competition. Dr. David Wise was the major professor under whom Dr. Johnson's thesis work was done. The Doctoral Dissertation Award Committee of ACM recommended its publication to expedite the marriage, in practice, of software methodology to VLSI circuit design. The concepts of "functional programming" in applicative languages to achieve an effective "silicon compiler" are clearly and forcefully articulated.

Walter M. Carlson
Chairman, ACM Awards Committee

Acknowledgments

I would like to thank Daniel Friedman and David Wise for their guidance and support throughout a most interesting course of study. They have shared both knowledge and learning with me, which is to say they are good teachers. I am grateful.

In the past year, David's infectious enthusiasm and ability to nurture emerging themes have been crucial to the progress of this investigation. He has skillfully worn all the hats of a research director, but I value beyond that his friendship and patient encouragement.

John O'Donnell, Franklin Prosser, and Mitchell Wand contributed a diversity of perspectives on the subject I chose to investigate. I hope that each is reflected here in a fair light. I thank my committee for their attentive reading of the drafts of this work and apologize to them for any remaining errors.

The programming language presented herein was implemented by Anne Kohlstaedt and myself, between September 1980 and June 1982. Its development continues. We gained much from earlier implementations to which Cynthia Brown and Casper Martin contributed.

I am indebted to my parents, Anne and Dexter, for bequeathing to me a taste for knowledge and providing for me the opportunity to pursue it.

It is somehow saddening to write that Jennifer Deam, who devoted the same portion of life as I did to this endeavor, must now make do with a brief acknowledgment. If her contributions to this work, and presence in it, are intangible to its readers, it is comforting to know that she shares the joy of its completion. And so, to Jennifer, with love, I thank you.

Preface

The discipline of *applicative* program design style is adapted to the design of *digital synchronous* systems. The result is a powerful and natural methodology for engineering correct hardware implementations. This book presents a method to develop digital system descriptions from recursive specifications; offers a prototype general purpose modeling language that supports this design task; and makes a formal connection between functional recursion and component connectivity that is pleasantly direct, suggesting that applicative notation is the appropriate basis for digital design.

Design is a translation of notation from an abstractly descriptive *specification* to a concretely descriptive *realization*. Recursion equations are used as a specification language. The realization language is another form of recursion in which variables denote sequences (rather than functions) that represent digital component behavior. Self-reference in realizations corresponds to feedback in a physical implementation.

Synthesis is a method for constructing realizations that are guaranteed to meet their specifications. It is a synonym of "engineering" peculiar to computer science, where the concern is not only with methods but also with their automation. This term suggests a factor of human guidance, as opposed to *compilation* which does not. Realizations can, however, be compiled from *iterative* specifications. Even for the case of non-iterative specifications, synthesis of an iterative version is the primary tactic here. This tactic formalizes the conventional digital design technique of decomposing a circuit into an architecture and a finite state controller.

The formal setting for a discussion of this topic is the calculus of Scott and Strachey. A specification denotes the fixed point of a functional; a realization denotes a fixed point in a domain of sequences. This approach to synthesis, then, is yet another application of modeling "function recursion" with "data recursion", or *reflexivity*. An interpreter has been implemented for *DAISY,* a dialect of the Scott-Strachey notation. Any description expressed in DAISY can be directly executed at successive steps in its evolution. Thus, the notation that serves as the medium of engineering serves also as a vehicle for experimentation. This is important to the practice of design because the engineer can explore some aspects of performance without expensive constructions of hardware prototypes, or risky recodings in a simulation language.

Two examples follow. A non-trivial exercise in language-driven design, derivation of a controller for an applicative language interpreter, reveals that powerful global structuring techniques, such as hierarchical decomposition and data abstraction, are inherited immediately from the functional style of description. Executability of the current description at each stage of the derivation provides a model for testing representation decisions and trivial modifications. Next, a specialized algebra is developed to address a typical local refinement problem: reducing external connections by means of serialization. Thus, local as well as global design problems yield to the applicative method.

Applicative notation is especially suitable for digital circuit description because the basic algebra is the same in both realms. Even though the underlying symbols are interpreted differently (*i.e.* operations *vs.* components; values *vs.* signals) the manner of combining them (*e.g.* composition/construction *vs.* serial/parallel wiring) is identical. Hence, recursion equations, McCarthy's *mathematical* basis for the science of computation, is fitting for hardware design because it so well reflects the *physical* basis of computation: digital electronics.

Contents

1. Introduction   1

  1.1. Summary   4

  1.2. Related Research   7

    1.2.1. Sequential Formal Models   8

    1.2.2. Operational Aspects of Modeling   11

    1.2.3. Other Motivations   11

  1.3. Outline of the Presentation   12

2. The Specification Language   15

  2.1. Typed Recursion Equations   16

  2.2. Solutions of Specifications   21

  2.3. Reasoning about Specifications   24

    2.3.1. Structural Induction   24

    2.3.2. Subgoal Induction   25

  2.4. Transformations on Recursion Equations   28

    2.4.1. Grammatical Transformations   29

    2.4.2. Distributivity of the Conditional and Multiplexors   30

    2.4.3. Combined Operations   31

    2.4.4. Universal Schemes   33

    2.4.5. Synthesis of Iterative Form   37

  2.5. The Scott-Strachey Notation   40

    2.5.1. Flat Domains   41

    2.5.2. Non-flat Domains   42

    2.5.3. Domain Operations   43

    2.5.4. Functionals   44

    2.5.5. Recursion   44

    2.5.6. Reflexivity   45

  2.6. Other Issues   45

    2.6.1. Specifying the Specification Language   45

2.6.2. Specifying Control	46
2.6.3. Distributivity of the Conditional, Revisited	48
2.6.4. Multiple Valued Functions	49
3. The Realization Language	**51**
3.1. Digital Circuit Descriptions	52
3.2. Translation to Circuit-Description Form	54
3.3. Decomposition of Combined Components	57
3.4. Circuit Synthesis	61
3.5. A Domain Model of Behavior	63
4. DAISY	**67**
4.1. Operational Semantics – a Summary	67
4.2. The Language	69
4.3. Formal Semantics of a Subset of DAISY	71
4.4. Circuit Emulation	80
4.4.1. Non-finite Data Structures	80
4.4.2. Output Driven Computation	82
4.4.3. Experimentation with Realizations	83
5. Design Examples	**89**
5.1. Higher Level Components	90
5.1.1. Packaged Combinations	92
5.1.2. Abstract Components	93
5.2. Language Driven Design – Introduction	101
5.3. Application to Language Driven Design	102
5.3.1. The Language L	103
5.3.2. An L-interpreter	105
5.3.3. Definition of IM	107
5.3.4. Stacking Version of IM	112
5.3.5. Simple Loop for the L-interpreter	114
5.3.6. Some Refinements in the Loop Version	114
5.3.7. Realization of IM	116
5.3.8. Refined Realization of IM	116
5.3.9. Remarks	121
6. Circuit Refinement	**123**
6.1. Transformation Axioms	126
6.2. General Transformations and their Behavioral Interpretation	128

6.3. Scheduling Derivations   131

   6.3.1. Circuit F   132

   6.3.2. Circuit G   133

   6.3.3. Circuit H   136

6.4. Remarks   139

7. Conclusion   141

7.1. Review   141

   7.1.1. Iteration   141

   7.1.2. Circuit Synthesis   142

   7.1.3. Circuit Refinement   143

7.2. Limitations of the Approach   143

   7.2.1. Bidirectionality   144

   7.2.2. Digital Asynchrony, Communication, and Integration   144

7.3. Prospects for Research   147

   7.3.1. Multiphase Clocking   147

   7.3.2. The Realization Language as a Formal System   148

   7.3.3. Other Topics   148

7.4. Final Remarks   149

Selected Bibliography   151

Appendix

A. True Syntax of DAISY   159

B. DAISY Trials   163

C. Proofs   191

D. Table of Symbols   201

Index   203

Figures

2.1 A Standard Semantics for Terminal Terms   47

4.1 DAISY Expression Syntax   70

4.2 DAISY's Kernel Syntax   74

4.3 Conversions to the Kernel Language   75

4.4 DAISY's Standard Semantics

    a. Domains   76

    b. Valuation   77

    c. Auxiliaries   78

4.5 Some DAISY Operations   79

4.6 DAISY Component Implementations   82

4.7 Experiment with the FAC Realization   86

4.8 Experiment with the FIB Realization   87

4.9 Experiment with the GCD Realization   88

5.1 A Schematic for Circuit C   96

5.2 Experiment with C_{FIB}

    a. Source for the Realization   99

    b. Record of an Experiment   100

5.3 Standard Semantics of the Language L   104

5.4 Non-linear Specification for an L-interpreter   111

5.5 Stacking Version of the L-interpreter   113

5.6 Simple Loop for the L-interpreter   115

5.7 Refined Loop for the L-interpreter   117

5.8 Higher Level Components for the L-realization   118

5.9 Realization of the L-interpreter   119

5.10 Refined L-realization   120

5.11 Continuation Semantics for L   122

A.1 Present DAISY Syntax   159

A.2 Conversions to Present DAISY Syntax   160

Synthesis of Digital Designs
from Recursion Equations

1. Introduction

Advocates of *applicative* programming style claim that it is somehow closer to the intuitive process of conceiving an algorithm and is, therefore, the proper notation for the development of computations. Since few deny the need for better programming methods, this "applicative premise" has received a good deal of scrutiny over the past twenty years. Much of this research demonstrates that the approach is viable; that is, it shows that the discipline can be used successfully to attack problem classes that at first glance appear to be beyond its capabilities.

The research reported here began as a test of the applicative premise in a fundamental and difficult problem area: the design of hardware. The original approach was to make an earnest attempt at denial of the premise by showing that it was not an appropriate basis for addressing the problem of circuit description. The attempt failed, for I found that a purely functional notation is quite viable for *digital* circuit design, and is in some ways preferable to conventional engineering. The substance of this investigation lies in the design method that evolves from strict adherence to applicative style.

The main conclusion here is that hardware designers can be comfortable with this method because they have been thinking applicatively all along. By adopting a digital implementation technology the designer orients a circuit in time and thinks of it as a function (on state) rather than as a feedback system in equilibrium. This temporal constraint on product behavior reflects an abstraction from, and a simplification of, the physical elements of electronics. The abstraction is made in order to attain a tractable intellectual basis for organizing behavior. Of course, abstraction is a quality of any design discipline; but the correlation between the motives of *digital* design and *applicative* style is not merely superficial; the *means* of abstraction, functionality, is the same.

Automation of circuit design—and automation is an eventual goal of this work—entails finding a representation for circuits. This appears to be a profound obstacle to applicative style, for it necessitates building a data structure that describes *feedback,* a manifestly circular physical quality. It is not immediately obvious how to construct circular data in a notation that prohibits expressed side effects. The solution is to use recursion (*i.e.* reflexivity through fixed point constructions) to describe connectivity. In doing so, one simultaneously obtains a description of the product *and* a model of its behavior.

Design is viewed as a translation of notation, starting with a *specification* and ending with a *realization*. The specification language should be abstractly descriptive; its main purpose is to convey thought. The realization language should be concretely descriptive in the sense that it portrays an implementation accurately enough to serve as a starting point for fabrication. Specifications will be systems of *typed recursion equations* expressed in the style of McCarthy (1963). The realization language is a linear form of circuit schematic in which the connectivity of the circuit's components is expressed by equation rather than a drawing of boxes and lines.

Manna defines *synthesis* to be "the theory for constructing programs [read: realizations] that are guaranteed to be correct [with respect to their specifications] and therefore do not require debugging or verification" (1974, p. 219). It is a synonym for *engineering* that is peculiar to Computer Science, which is concerned not only with methods but also their automation. It is used here to suggest a system that employs human guidance to produce realizations (as opposed to *compilation,* which does not). However, I do not present a mechanized system here; synthesis is carried out by hand, although some of the steps clearly can be automated. An essential feature of synthesis is that the meaning of the specification is preserved, or at worst, altered in a perceptible way. For example, realizations may be acceptable even though they are only *partially correct*; that is, they produce the correct answer whenever they "halt" but sometimes diverge when the specification does not. The designer may prefer to strengthen the specification, rather than reject the realization, according to some ulterior motive (such as a presupposed architecture). The process of synthesis gives a context for this kind of decision making.

The method of synthesis used here is *transformation*. A collection of correctness preserving rewriting rules is used to derive realizations. Burstall and Darlington (1977, p. 46) characterize this method as "an inference system in which the sentences are recursion equations." Transformation is a relatively direct form of synthesis, a formalization of step-wise refinement. Other forms generate realizations as a byproduct of some other activity, such as the proof of a theorem[1] (Manna and Waldinger, 1971, 1979).

The conventional approach to digital design centers on the development of a sequential algorithm to control the architecture of the circuit. The controller is usually presented as a finite state machine, flowchart, or imperative (*i.e.* statement-oriented) program. Various methods exist to translate this abstraction of control into hardware. The control algorithm serves as a basis for making representation decisions about the architecture. The approach presented here is fundamentally similar although it is carried out in a functional notation. The initial objective of transformation is to find a version of the specification that is in iterative form. From there, construction of a circuit description is straightforward. Since iterative form characterizes "flowchartability," its synthesis has been studied as a means to derive programs. The fruits of this research are directly applicable to circuit synthesis.

Use of a functional specification style must be justified partly as a matter of preference. It is an attempt to cast design in a clean mathematical setting. If, as is the case here, the principal goal is correctness, then an unambiguous meaning for source and target notations is a necessary starting point. Functional style has additional advantages as a basis for digital hardware design. The functional programmer and the digital designer have similar vocabularies. The ubiquitousness of the word "function" in their discourse is testimony to that similarity. To the programmer however, "function" is more a noun; to the circuit designer it is more a verb. The programmer deals with operations and values; the designer deals with components and signals. In the latter case a notion of orderly activity over time is implicit: a component *behaves*. That is, the programmer and the designer think in the same sentences but with a different semantics. Nevertheless, they use the same algebra, manipulating basic symbols by such rules as

[1]What constitutes a specification or realization is relative. In the synthesis system cited here, recursion equations serve as the realization language.

composition (wiring in series), structural combination (wiring in parallel), and the use of selection (multiplexing) to achieve a function (functioning). I exploit this commonality to achieve a basis for design.

Specification and realization meanings are unified in the functional programmer's metalanguage: the calculus of Scott and Strachey. This language has a computable semantics, and an interpreter is presented for a crude dialect called *DAISY*. DAISY could serve as a medium for transformation, but its role in this investigation is only as a vehicle for experimentation. Direct executability of the descriptive notation has three practical benefits. The evolving design can be demonstrated without prototyping or translation to a simulation language. Gaps in the automation of synthesis can be bridged empirically (Here the gap is quite large since none of the synthesis is yet automated). Most important, emulation of target behavior can reveal properties of performance that are not addressed formally, either because they are not describable in the specification language or because they are not worth establishing through formal means.

The early chapters that follow make the formal connection between specifications and realizations. Later, this foundation is applied to two non-trivial design tasks: a controller for a programming language interpreter is synthesized, and a specialized transformation system is defined to address a problem in local circuit refinement. It is shown that various standard structured design techniques are reflected naturally in this methodology (Section 1.3 expands this prospectus).

1.1. Summary

The following expressions are substantially equivalent:

$$F(a) \text{ where } F(x) \Leftarrow p(x) \rightarrow f(x), F(g(x)). \qquad (S)$$

$$
\begin{aligned}
&x := a \, ; \\
&\text{while } \neg p(x) \text{ do } x := g(x); \qquad (P) \\
&x := f(x)
\end{aligned}
$$

$$
\begin{aligned}
X &= \quad a \, ! \, g(X) \\
READY &= \qquad p(X) \qquad\qquad (C) \\
VALUE &= \qquad f(X)
\end{aligned}
$$

Specification S defines its value as the output of a recursive function F, whose defining equation has the structure generally associated with the program P. The system C is a linear description of the schematic[2] for the register transfer circuit

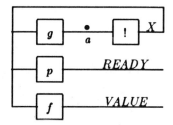

The component $\boxed{!}$ is a clocked storage element that for brevity I shall call a *register*. A synchronizing signal is common to all registers and, like the power supply, is suppressed in the schematic. The token "•" indicates that the register has been initialized with the value a, or more generally, that the circuit is now in a state in which its register contains a. In S and P the ground symbols p, f, and g are primitive operations, part of the vocabulary of a fixed (but often not otherwise specified) *underlying type* for making specifications. The circuit components $\boxed{g}$, $\boxed{p}$, and $\boxed{f}$ are counterparts of operations, but it is understood that they operate over time, continually producing a value that is a function of their present input. The register synchronizes the system and makes it possible to assume that component behaviors are discrete.

Each of S, P, and C is a canonical representative of its realm. If the underlying type is powerful enough, any partial recursive function can be transformed to a single repetition, and any program can be expressed as a loop[3]. A schematic similar to C often accompanies an introduction to digital/synchronous systems (*e.g.* Mead and Conway, 1980, pg. 221; Hill and Peterson, 1968, pg. 250). It exhibits the characteristic property that all closed signal paths pass through a register.

---

[2]Left-right flow is used in schematics wherever possible, so that a component's inputs are on the left. This is informal notation and is accompanied by systems like C that state the input-output relationship explicitly.

[3]The heredity of this "folk theorem" is explored by Heral (1980). S is essentially Kleene normal form (Kleene, 1950, pg. 288), although the use of repetition rather than minimization suggests (Brainerd and Landweber, 1974, Corollary 5.7).

The correspondence between P and C is the basis for conventional structured hardware design. A standard formalism for describing a flowchart schema is to define the *value history* of its state (*e.g.* Manna, 1974; Greibach, 1975). The history is the sequence of values the state will acquire, expressed as a first order linear recurrence relation depending on the current state and the current label (which can be given as part of the history). If the monolithic state history is decomposed into individual variable histories the result is a simultaneous first order recurrence relation. It is also a register transfer description, describing how each stored value in the circuit will change as a function the present register content.

My approach is to derive individual histories from (systems of) recursion equations rather than from flowcharts. Since an essential step in the process is to place the equations into iterative form, the passage from specification to realization will only subliminally construct a flowchart. In the circuit description C above, the notation suppresses the recurrence. This is valid because the dependency is fixed by the nature of components, and worthwhile because a concise description of connectivity emerges. The crucial transition between specification and realization *notations* changes the interpretation of the primitive symbols. I refer to this transition in meaning as *lifting*.

The formal connection between source and target languages is made on the basis of the forms S and C above. Consequently, synthesis decomposes into a subtask of transforming an initial specification to an instance of S, followed by a sequence of refinements to the corresponding instance of C. I shall show that much of the transformational algebra used to obtain S is transparent to (*i.e.* distributes over) lifting. A less succinct class of specifications, called *simple loops*, are immediately realizable and yield more informative schematics.

In theory any system of recursion equations can be expressed as a simple loop, but to obtain it one must assume that the underlying operations are powerful enough to implement recursion. For example, if arithmetic is available, an encoding can be used to represent a control stack. If such computational power is not assumed—as may well be the case in digital design—then not all specifications can be directly realized. Varying the rules about what is admissible as an operation induces a complex hierarchy of transformability on the class of all specifications. Under a minimal set of assumptions simple loops can be constructed from any *linear* specification (*i.e.* one without nested recursive calls).

Recursive linearity characterizes flowchartability, and translation of recursion equations into flowcharts has been widely studied. The results are all applicable to my principal strategy of circuit design.

Separation of control and representation is a common theme in algorithmic design, independent of the implementation realm. Compare Hoare's remark on structured programming (1972)

> In the development of programs by stepwise refinement ..., the programmer is encouraged to postpone the decision on the representation of his data until after he has designed his algorithm, and has expressed it as an 'abstract' program operating on his 'abstract' data. He then chooses for the abstract data some convenient and efficient concrete representation in the store of a computer; and finally programs the primitive operation required by his abstract program in terms of this concrete representation.

to that of Winkel and Prosser concerning structured digital design (1980, p. 131).

> One of the first steps of a top-down design is to partition the design into (a) a *control algorithm* and (b) an *architecture* that will be controlled by this algorithm. The top-down analysis will suggest a rough preliminary version of the system architecture, involving abstract building blocks such a registers, memories, and data paths...
>
> Next, we work out the details of the control algorithm at an *abstract level*. The control algorithm is in many cases surprisingly independent of the hardware...

In the functional specification style, this separation of aspect entails postulating a type on which the specification will operate. To proceed in this way it is clearly desirable to start in a fully abstract setting, but with the knowledge that once a type emerges it is certain to have a computable representation. A fully abstract functional language exists in the notation of the Scott-Strachey calculus (Stoy, 1977). In addition to serving as a most-abstract starting point for design, the Scott-Strachey calculus is also used as a metalanguage through which the meanings for specifications and realizations are unified. Through interpretation the metalanguage also serves as a vehicle for experimentation.

1.2. Related Research

The fact that the formal language used for mathematical models also has a computable meaning leads to some difficulty in classifying the bearing of other research on this presentation. It makes the useful distinction between formal and operational models somewhat hazy, since there is that sense in which a mathematical formulation can be considered a program. The confusion is nowhere better illustrated than in the vanguard work of McCarthy, a founder of the functional specification style, who proposed using recursion equations as a basis for the specification of computations (1963) and simultaneously provided an interpreted language, LISP (McCarthy, *et.al.* 1965), for experimentation. Both of McCarthy's contributions are influential to this investigation. The problem is that LISP's interpreter is often confused with the underlying mathematics.[4] Thus, the division of the review below into formal models and operational aspects of modeling is for the most part artificial. Most of the researchers cited are involved in both areas.

1.2.1. Sequential Formal Models.

In Chapter 3 a notation is defined for the description of digital behavior. Two equivalent definitions are given for the meaning of the notation: the first associates each ground symbol with a function on the natural numbers; in the second definition each symbol denotes a sequence. The first definition should be familiar to hardware designers, for it corresponds to the usual interpretation of a circuit's state as a first order linear recurrence relation. A restatement of the model in terms of value histories gives a domain formulation of the same model. It is essentially the same formulation as Kahn's (1973) and was foreseen as early as 1965 by Landin. It is introduced as a prelude to the implementation of the model in Chapter 4.

Any characterization of computation in terms of discrete value histories can be construed as an approach to digital design. Extraction of histories is frequently used to formalize or analyze programming constructs. Texts by Manna (1974) and lecture notes by Greibach (1975) employ this approach to describe flowchart schema. Each goes on to develop formally the relationship, first noted by McCarthy (1963), between flowchart and recursion schema. They therefore establish, albeit indirectly, the basis of my approach to digital design.

The symbolic evaluator of Cheatham, Holloway, and Townley (1979) derives "the recurrence relations that describe the behavior of loop variables," as a means for the analysis and verification of imperative programs; but they have at

---

[4]Stoy (1977, p. 182) gives a very clear discussion of this point.

the same time produced a digital circuit assembler and optimizer according to my model.

The notion that functional style employs same algebraic framework as digital designer is perhaps best illustrated in the FP programming movement (Backus, 1978, 1981a, 1981b). An algorithm can be expressed in a purely combinatorial form that corresponds to how circuits are physically wired. The FP style goes beyond my goal however, by promoting variable-free programming[5], and thereby suppressing a quality that I shall eventually emphasize: state. By suppressing state, one rids one's self of a mathematically clumsy concept. However, FP languages invariably have a construct for iteration (*e.g.* Backus's *insert* operator) and therefore retain the computationally necessary concept of an accumulator. Structured digital design methods usually focus on accumulator behavior; hence, accumulator outputs are given names. The identifiers correspond exactly to the program variables of a recursive specification.

In his dissertation, Cohen (1980) also uses an iteration construct as a basis for transforming recursion equations into programs. He also gives a fairly thorough review of research in the compilation of recursion equations. The circuit synthesis techniques presented here extend many of these results to a different implementation realm.

Iterators often carry an implicit termination condition, reflecting the programmer's preoccupation with that property. However, termination is not a quality enjoyed by circuits if they are modeled in terms of their temporal behavior. The more natural abstraction is that of infinite behavior which occasionally notifies the outside world that meaningful events are taking place. Ashcroft and Wadge (1977) present a formal system called LUCID in which non-finite histories are implicit. LUCID is presented as a programming language that incorporates iteration in a "mathematically respectable way." The circuit description C, above, is easily recognizable as a LUCID program, and in fact has the appropriate semantics. The use of circuit description text as a formal system to support inference is briefly discussed in Chapter 7. In his dissertation, Meyers (1980) also investigates the use of non-finite structures in programming. While the

---

[5]Backus does not prohibit the use of variables in expressing algorithms, but seeks to reduce their influence on conceiving algorithms. See (Backus, 1981).

properties of non-finite, especially sequential, objects are central to the development below, their use as a *programming* construct is not. They are used instead to model the properties of electronic components and thus serve here as a target language. At the same time however, the circuit descriptions are legitimate applicative programs. The implications for programming style are not explored.

The preferred formal setting is the functional calculus of Scott and Strachey. A domain formulation of behavior, presented here in Section 3.5, here was proposed by Kahn (1973, Kahn and MacQueen, 1977). The set of *signals* over a set of primitive values is defined by the domain isomorphism

$$Signal = Value \times Signal$$

(Kahn uses the domain of sequences, $Value^\omega$, which is the essentially the same domain). That is, a signal is an infinite sequence of values. Components are processes that produce and consume signals.

Milner has developed a robust mathematical foundation for describing process semantics (Milner, 1973, 1980a; Milne and Milner, 1979; Gordon, 1980), in which my model can easily be embedded. He characterizes process behavior as a point in the domain

$$Behavior = Input \rightarrow (Output \times Behavior)$$

A *component* gives rise to a function from signals to signals. Given a process behavior and an input signal it is a trivial coercion to construct the right output signal. The basic difference here is that components are defined as higher order signals; that is, as sequences of operations. Application is generalized to deliver the induced signal-to-signal function. This is merely a technical adjustment in light of the fact that the only components I will allow are constant sequences. This constraint is *temporarily* relaxed in Chapter 5 to introduce communication.

Gordon (1981a, 1981b), Cardelli (1980, 1982), and Milner (1980b) use process semantics in microcosm to describe circuit behavior. They develop a mathematically attractive notation for circuit analysis and verification. For this purpose their notation is superior to the applicative notation used here because it appears to address a wider class of circuits. However, the goal here is synthesis, and a purely applicative target language is sufficient to realize purely functional specifications. We return to this point in the conclusion.

1.2.2. Operational Aspects of Modeling.

My approach to synthesis maps (fixed points in) a domain of functionals to (fixed points in) a domain of signals; it takes self-reference in the guise of recursion to self-reference in the guise of feedback. Implementing recursion with "reflexivity" is commonplace in programming. Compilers use program pointers to manage control; reduction interpreters use shared text to optimize substitution.

More overt forms of *data recursion* are often presented as advanced functional programming techniques (*e.g.* Friedman, Wise, and Wand, 1976; Burge 1975; Henderson, 1980). The earliest example is Landin's use of *streams* in conjunction with his effort to give an applicative operational description of ALGOL 60 (Landin, 1965). It is worth noting that he introduced streams in order to factor (index) variable histories out of loop statements, but immediately observed that the same mechanism "would be used to model input-output if ALGOL 60 included such". He elected to represent histories as lists, and had to confront the possibility that non-terminating loops would produce infinite histories. He could not directly express infinite data structures in his "call-by-value" modeling language, and used function closures as a delay mechanism to defer the possibly divergent, and anyway untimely, construction.

In 1976, Friedman and Wise proposed that this closure trick be incorporated into the primitive data space operations so that all computation is deferred until it becomes timely. A similar mechanism was independently presented by Henderson and Morris (1976), and suggested earlier by Vuillimen (1974) and Wadsworth (1971). The effect on a conventional reduction interpreter is profound, for a *suspending* constructor induces an outermost reduction rule. Under reasonable assumptions about the underlying operations, outermost reduction is consistent with the formal meaning of an expression as a least fixed point. Moreover, non-finite data structures can be built and manipulated as a matter of course; constructs like Landin's streams become transparent. The interpreter for DAISY is implemented on a virtual list multiprocessor that uses suspending construction. It is therefore possible to express specifications and realizations without fear that they will be compromised by an overly strict interpreter.

1.2.3. Other Motivations.

Between 1976 and 1980 Friedman and Wise published several articles (1976c, 1977, 1978a, 1978b, 1979) promoting a purely applicative specification style and showing that it could be applied to "systems programming" problems. Since a circuit is a system, it seemed evident that the approaches they were suggesting would be a possible basis for hardware design.

I owe much to Wand's work in compiler generation (1980a, 1980b, 1982a), as can be seen from the choice of example in Chapter 5, and the style in which it is developed. He gives a decidedly small set of generalized combinators that captures the code-structure of conventional programming languages, and develops a formidable strategy to decompose a semantic definition into a compiler/machine pair. The machine "factor" is in iterative form, and it follows from this investigation that it can be used to construct special purpose hardware for the direct execution of compiled code. In the example I develop, however, Wand's elegant factorization is omitted since its goal is not really at issue here; a direct interpreter is derived instead. The reader who is uncomfortable with the resulting machine is urged consult Wand's work for insights into how I might have arrived at a more conventional implementation.

1.3. Outline of the Presentation

Chapter 2 reviews the language of typed recursion equations that I refer to as *specifications*. Basic methods for reasoning in and about this language are reviewed. The chapter serves not only to state preliminary results, but also to give an introduction to readers who are unfamiliar with the description style. Three examples, representing iterative, linear, and non-linear specifications, are presented and subsequently used to follow the development through Chapters 3, 4, and 5. A series of extensions to the specification language are made, starting with the incorporation of structural combination and a selection primitive, and ending with the admission of stacks to the underlying type as a means to implement recursion. The extensions make it possible to transform various kinds of specifications into *simple loops*. The final sections of the chapter review the notation of the Scott-Strachey calculus, which is used to address issues that arise later in the presentation. Among the issues discussed are the specification of semantics, which will be the starting point for a lengthy synthesis exercise in Chapter 5; and the use of continuations to specify control.

Chapter 3 defines a realization language for describing the logical behavior of digital circuits, and makes the fundamental connection to the specification language. It is then established that the functionals used to combine operators may also be used to combine components. As a result, simple loops are shown to be essentially realizations, lacking only a lifting of the interpretation of ground symbols. The digital model is restated in the terms of the Scott-Strachey calculus, as a prelude to an implementation of the model in DAISY.

A direct semantics for DAISY is defined in Chapter 4, along with a brief summary of its implementation. Basic programming techniques for circuit experimentation are defined. The chapter concludes with a series of experiments on the example specifications. In one case, observation of the derived circuit's behavior reveals an interesting property of performance that is not addressed in the specification.

Most of Chapter 5 is devoted to a non-trivial design exercise: the synthesis of an interpreter-circuit for an applicative programming language called *L*. To attack larger design tasks, we must of course adopt structured design techniques. The transparency of structural combination to lifting makes possible the hierarchical decomposition of realizations into *packaged components*, the behavioral analog to the programmer's "macro". The technique of information hiding also lifts, resulting in a factorization of *abstract components*. This decomposition leaves a residue signal of *instructions*, and forces us to confront the issue of overt communication for the first time.

The *L*-interpreter's derivation begins with a formal definition of the language, which is a non-linear, fully abstract specification. Of the six major steps in the transformation, two require substantial designer creativity. The first step is to propose a more concrete specification of *L* and hence is mainly concerned with finding an underlying type for interpretation. Once a type has been adopted, construction of a simple loop version of *L* is straightforward, although to reach a linear version some control decisions must be made. I pause to do some register optimization, presented as a creative task. An improved loop is transcribed to a circuit description, from which abstract components are then extracted. In Appendix B, the successive descriptions are given in DAISY and executed to show the logical behavior of the evolving design.

Chapter 6 suggests an approach to circuit refinement. A specialized set of transformation rules is tailored to address a complexity problem in large scale design. The task is to "fold" a combinatorial system with many external connections into a synchronous system in which computation is serialized. The derived circuit is a data-flow element in which some of the connective storage is realized.

The transformation process produces as a byproduct a computation schedule that can be used to coordinate the refined circuit with the surrounding computation.

Chapter 7 reviews the presentation, discusses some of its shortcomings, and suggests areas for further investigation.

The language DAISY is presented in a somewhat idealized form in Chapter 4. Appendix A gives the present syntax. Appendix B shows the DAISY source for running examples throughout the presentation. Appendix C contains proofs of some of the propositions in the body of the dissertation. Appendix D is a table of symbols.

The primary motive of this study was to extend McCarthy's "mathematical basis for the science of computation" (1963a, 1963b) in the direction of its physical basis. This area is an excellent test bed for the discipline of applicative style, but its goal should not be taken as the description of all hardware. In electing digital implementation technologies hardware designers have already adopted functionality as their fundamental abstraction and can profit further from a design methodology that stems from the same foundation. It is my hope that those familiar with conventional digital design methods will see in this presentation a fitting basis for their craft. However, the profound formal foundation and rich notation that have evolved from McCarthy's basis can be a hindrance. It is hardly reasonable to expect the "uninitiated" reader to absorb all the principles without first perceiving a payoff. The reader who is unacquainted with functional style should consider reading this material in two passes, first to see its direction and then to fill in the details. On first reading, one might do well to skip Sections 2.5, 2.6, 3.4, 4.3, 5.3.2, and 5.3.3, for it is in these sections that I formally address issues that are either on the fringe of the subject at hand or are intuitive to anyone already familiar with the design of computations. A condensed treatment of the topics in Chapters 3, 4, and 5 can be found in (Johnson, 1984).

2. The Specification Language

A *recursion equation* is an equation whose variables range over functions. A *specification* is a system of recursion equations. Any specification has a canonical solution; it is the unique set of minimally defined functions that simultaneously satisfy the definitions. Hence, the specification language is unambiguous. This chapter reviews basic facts about recursion equations and ways to reason about and manipulate them. A thread of facts is established that leads to a connection with the realization language defined in Chapter 3. The thread unwinds through a sequence of extensions to the notation, making it possible to transform larger and larger classes of specifications into *iterative form*. Iterative form is a characterization of "sequential control", and thus coincides with the class of specifications that, under a minimal set of assumptions, can be associated with a flowchart description of a computation. Just as flowcharts are a frequently used basis for digital design, iterative specifications are so used here.

Specifications are made in terms of an *underlying type*, a collection of ground symbols that denote values and operations from which more complicated things are built. It is the designer's "implementation realm". If the realm is TTL logic, for example, the underlying type would have two voltage levels and a parts catalog of components. In practice, specifications will always be typed. However, some transformations on specifications are valid no matter what the underlying type is. Hence, we shall often be dealing with recursion *schemes* or recursion equations over uninterpreted ground symbols. Finding generally valid transformations is obviously desirable, since they are applicable in any realm.

There exists a "universal type" in which all others can be embedded. A rather sophisticated notation, called the *Scott-Strachey* language, has evolved around its use. It represents one limit to which the specification language might be extended. The language allows for the description of highly abstract entities, such as function-valued functions, and is in a sense too abstract for the purpose of synthesis. In using Scott-Strachey notation as a starting point for a design, the first step will always be to propose a representation—that is, a suitable underlying type—for a more concrete specification.

McCarthy is generally acknowledged as a founder of the functional specification style. The syntax of the specification language is similar to the language he uses in early articles (McCarthy, 1963a, 1963b). Much of the our basic vocabulary comes from an introductory text by Wand (1980). Manna's text (1974) and Griebach's lecture notes (1975) are good introductions to the relationship between recursion equations and flowchart schema. Both cite the landmark works in this area.

2.1. Typed Recursion Equations

A design will be implemented from basic components, and in making a specification, this vocabulary is usually fixed in advance. This set of "building blocks" is called the *underlying type* of the specification.

DEFINITION 2.1-1. A type **D** *consists of*

    i.   *A* carrier *set, D, of values.*

    ii.   *A set of* constants, $C \subseteq D$.

    iii. *A finite set of total* operations, $f:D^n \to D$ *for various n.*

    iv. *A finite set of total* predicates, $p:D^n \to \{\text{true, false}\}$, *for various n.*

An operation $f:D^n \to D$ is said to be an *n-place* operation. C and D are often equal, but when containment is proper, D will always be inductively defined from C. That is, D will be the smallest set containing C and closed under the operations of **D**. An *indeterminate* constant ϕ is sometimes appended to C, and the set of truth values may likewise be extended. Depending on the context, ϕ is either unknown (don't-know) or its value doesn't matter (don't-care). An example of a type is **Dig**, for digital logic, with carrier $Dig = \{ high, low \}$, 2-place operation

$$nand(x,\ y) = \begin{cases} low \text{ if } x = y = high \\ high \text{ otherwise} \end{cases}$$

and 2-place predicate $high?$ = {$(high,\ true),\ (low,\ false)$}. **Dig** can be extended by introducing ϕ to its carrier[1]: $nand(\phi,\ y) = nand(x,\ \phi) = \phi$; and $high?(\phi) = \phi$.

Most of the examples in this chapter are arithmetic; they have underlying type **Int** of integers, with carrier Int = {...$-2,\ -1,\ 0,\ 1,\ 2,...$ }; constants Int; 1-place operations inc and dcr (increment and decrement); 2-place operations add, sub, mpy, and div (add, subtract, multiply and divide); 1-place predicate $zero?$ and 2-place predicates $lt?$ and $eq?$ (test for zero, less-than, and equal). **Int** is more primitively defined as having constant set {0 }, operations inc and dcr, and predicate $zero?$. Int is inductively defined as the smallest set containing {0 } that is also closed under inc and dcr.

A set of symbols is associated with the underlying type and serves to represent it in the specification language. When it is necessary to make a distinction between symbols and their abstract counterparts, symbols are either underlined or enclosed in the quotation delimiters ' ⟦ ' and ' ⟧'. For **Int** the symbol set includes $\underline{inc}$, $\underline{dcr}$, $\underline{eq?}$, $\underline{\phi}$, $etc.$; and a numeral for each integer.

The letters u, v, w, x, y, and z are *identifiers;* they serve as formal parameters in function definitions.

Strings of upper case letters, such as *'FAC'* and *'GCD'*, are *function variable symbols,* which are defined by equation. The letters F, G and H are the function variable symbols usually used. The *rank* of F is the number of formal parameters it requires.

In discussions where the underlying type is not explicitly mentioned, the metalinguistic variables f, g, and h will range over operations; p and q will range over predicates; a, b, and c will range over constants; F, G, and H will serve as function variable symbols; and x_1, x_2, x_3,... will stand for identifiers.

Specifications are built from applicative expressions involving ground symbols and the special character set { ⌊ , ⌋ , ⊥ , ⇒ }

---

[1]Depending on the implementation technology, it may be more appropriate to define $nand(low,\ \phi) = nand(\phi,\ low) = high$ (Mead and Conway, 1980, p. 15).

DEFINITION 2.1-2. The language L_T of terminal terms *is defined inductively by:*

i. $c \in L_T$ *for constant c.*

ii. $x \in L_T$ *for identifier x.*

iii. *If f is an n-place operation and t_1, t_2 ,..., t_n are terminal terms, then*
$$f \lfloor t_1 \perp t_2 \perp \cdots \perp t_n \rfloor \in L_T.$$

The language L_R of recurrent terms *is defined inductively by:*

i. $L_T \subseteq L_R$

ii. *If F is a function variable symbol of rank n and t_1, t_2 ,..., t_n are recurrent terms, then $F \lfloor t_1 \perp t_2 \perp \cdots \perp t_n \rfloor \in L_R$.*

The language L_E of expressions *is defined inductively by:*

i. $L_R \subseteq L_E.$

ii. *If f is an n-place operator symbol and e_1, e_2 ,..., e_n are expressions, then*
$$f \lfloor e_1 \perp e_2 \perp \cdots \perp e_n \rfloor \in L_E.$$

iii. *If F is a function variable symbol of rank n and e_1, e_2 ,..., e_n are expressions, then $F \lfloor e_1 \perp e_2 \perp \cdots \perp e_n \rfloor \in L_E$.*

iv. *If p is an n-place predicate symbol and l, r, e_1, e_2 ,..., e_n are expressions, then the* conditional expression $p \lfloor e_1 \perp e_2 \perp \cdots \perp e_n \rfloor \rightrightarrows l \perp r$ $\in L_E.$

The substring to the left of the $\rightrightarrows$ in a conditional expression is called a *propositional expression*. Unqualified, the word "term" means "recurrent term". Our interests center on the function variable symbols and what they denote. Hence, they are called *serious* symbols; all other symbols are *trivial*. Terms and expressions inherit the qualities of their components.

DEFINITION 2.1-3.

A term (expression) is called serious *if it contains a function variable symbol. Otherwise, it is* trivial.

A term (expression) over identifiers x_1 ,..., x_n is one that contains no identifiers other than x_1, ..., x_n.

A ground *term (expression) is one that contains no identifiers.*

We can now define a *specification* to be a system of function-defining equations. The left-hand sides of these equations are "calling patterns" consisting of a function name and a formal parameter list. The right-hand sides are defining expressions, stating what the functions do when called. Two additional special symbols, $\Leftarrow$ and $.$ are needed.

DEFINITION 2.1-4. A recursion equation *has the form*

$$F \lfloor x_1 \lrcorner \cdots \lrcorner x_n \rfloor \Leftarrow e \div$$

where F is a function variable symbol of rank n, and e is a expression over x_1 ,..., x_n. This equation is said to be F's defining equation. A specification is a finite set of recursion equations, each defining a unique function variable symbol.

Note that the definition prohibits "global" identifiers. That is, a function's defining expression involves only identifiers in the function's parameter list. The following examples of specifications in **Int** will be used to illustrate the ideas of this chapter and the next[2].

---

[2]It is standard practice to switch to the more familiar infix notation for operations in **Int**. I will occasionally make the switch when doing so clarifies the presentation (in Section 2.3 for example). However, when making "official" specifications, I shall continue to use prefix notation, and beg the reader's indulgence, since in later chapters I would have to revert to prefix anyway.

$$GCD(x,\ y) \Longleftarrow eq?(x,\ y) \rightarrow x,$$
$$lt?(x,\ y) \rightarrow GCD(x,\ sub(y,\ x)),\ GCD(y,\ sub(x,\ y)). \qquad (S_1)$$

$$FAC(x) \Longleftarrow zero?(x) \rightarrow 1,\ mpy(x,\ FAC(dcr(x))\ ). \qquad (S_2)$$

$$FIB(x) \Longleftarrow lt?(x,\ 2) \rightarrow 1, add(\ FIB(dcr(dcr(x))),\ FIB(dcr(x))\ ). \qquad (S_3)$$

Intuitively, specification S_1 defines a *greatest common divisor* function, S_2 defines a *factorial* function, and the function defined by S_3 returns the xth element of the *Fibonacci* sequence: *1, 1, 2, 3, 5, 8....* . Two of these specifications are ambiguous; neither S_1 nor S_2 states what the function it is describing should return on a negative argument. This ambiguity will be resolved in the next section.

The three specifications differ in their structure. In S_1 the function variable symbols are outermost in all serious terms. In S_2 there is at most one function variable symbol in any recurrent term. S_3 has neither of these qualities.

DEFINITION 2.1-5. A recurrent term is linear *if it contains a single function variable symbol. A recurrent term is* iterative *if it is linear and its function variable symbol is left-most. A conditional expression "$p(t_1,\ t_2\ ,...,\ t_n\ ) \rightarrow r,\ s$" is linear (iterative) if each t_i is a terminal term and both of its branches, r and s, are either terminal or linear (iterative). A recursion equation is linear (iterative) if its defining expression is. A specification is linear (iterative) if each of its defining equations is.*

The recursive structure of a specification is of interest in itself, and will sometimes be considered independently of the underlying type. A *recursion scheme* is a recursion equation in which the ground symbols are left uninterpreted. A recursion scheme S' is called an *instance* of recursion scheme S if some or all of the uninterpreted symbols of S have been consistently replaced by specific symbols to get S'. For example, specification S_3 above is an instance of the nonlinear recursion scheme

$$F(x) \Longleftarrow p(x) \rightarrow c,\ h(\ F(g(g(x))),\ F(g(x))\ ).$$

2.2. Solutions of Specifications

The ground symbols in an expression denote the entities that they represent in the underlying type. Thus, the value assigned to any trivial term is simply the value of its abstract counterpart. Function variable symbols denote functions that *satisfy*, or are consistent with, their defining equations. To make the notion of consistency precise, we shall define a relation called *valuation* between specification text and meanings. Application of a serious function is interpreted as a textual replacement, called a *substitution*.

DEFINITION 2.2-1. Let e be an expression over identifiers $x_1, ..., x_n$ and let $t_1, ..., t_n$ be arbitrary expressions.

$$e \begin{bmatrix} t_1, ..., t_n \\ x_1, ..., x_n \end{bmatrix}$$

denotes the expression obtained by substituting t_i for each occurrence of x_i in e.

A specification gives a context for substitution in a valuation.

DEFINITION 2.2-2. Let S be a specification over a type with carrier D. Let c be a constant, f an n-place operator symbol, and p an m-place predicate symbol, denoting c^D, f^D, and p^D respectively. Let F be a function variable symbol defined in S by the equation "$F(x_1, ..., x_n) \Leftarrow \delta F$.", where δF is an expression. The function val *maps ground expressions to values in D as follows:*

$val \llbracket c \rrbracket = c^D$

$val \llbracket f \lfloor t_1 \lrcorner ... \lrcorner t_n \rfloor \rrbracket = f^D(val \llbracket t_1 \rrbracket, ..., val \llbracket t_n \rrbracket)$

$val \llbracket p \lfloor t_1 \lrcorner ... \lrcorner t_m \rightrightarrows r \lrcorner s \rfloor \rrbracket = \begin{cases} val \llbracket r \rrbracket, \text{ if } p^D(val \llbracket t_1 \rrbracket, ..., val \llbracket t_m \rrbracket) \text{ is true} \\ val \llbracket s \rrbracket, \text{ if } p^D(val \llbracket t_1 \rrbracket, ..., val \llbracket t_m \rrbracket) \text{ is false} \end{cases}$

$val \llbracket F \lfloor t_1 \lrcorner ... \lrcorner t_n \rfloor \rrbracket = val \llbracket \delta F \begin{bmatrix} t_1, ..., t_n \\ x_1, ..., x_n \end{bmatrix} \rrbracket$

The function *val* can be extended to a function over arbitrary expressions by providing an *environment* that gives values for free identifiers. That is, given a function $\rho : Ide \to D$, where *Ide* is the set of identifiers, add the clause "*val* $[\![x]\!] = \rho(x)$" to *val's* definition.

Given a specification, the value of a ground term can be derived by *reduction;* that is, through symbolic manipulation of the expression according to the rules of Definition 2.2-2. If a step in a reduction is justified by known properties of the underlying type, we shall call it a *simplification*. A step justified by the substitution rule is called an *unfolding* if the rule is applied from left to right. The inverse of unfolding is *folding*. We shall write "ΔF" to mean "by substitution according to F's defining equation" (unfolding), and "$\bigtriangledown F$" to mean "the abstraction of common subexpressions by identification, according to F's defining equation" (folding).

Recall the recursion equation

$$FAC(x) \Longleftarrow zero?(x) \to 1,\ mpy(\ x,\ FAC(dcr(x))\ ).$$

which we claimed earlier to specify a *factorial* function. Using Definition 2.2-2 and some simplification we can readily show that the expression *FAC(2)* reduces to *2:*

$FAC(2)\ =\ \underline{zero?(2) \to 1,\ mpy(2,\ FAC(dcr(2))}$	ΔFAC
$=\ \underline{mpy(2,\ FAC(dcr(2))}$	*conditional* $(\to)$
$=\ mpy(2,\ \underline{FAC(1)})$	*simplification*
$=\ mpy(2,\ \underline{zero?(1) \to 1,\ mpy(1,\ FAC(dcr(1))})$	ΔFAC
$=\ mpy(2,\ mpy(1,\ \underline{FAC(0)})$	$\to,$ *simplification*
$=\ mpy(2,\ mpy(1,\ \underline{zero?(0) \to 1,\ mpy(0,\ FAC(dcr(0))}))$	ΔFAC
$=\ mpy(2,\ mpy(1,\ 1))$	$\to$
$=\ 2$	*simplification*

Numerous mechanical steps have been omitted, as has any explicit mention of the valuation function. We simply allude to *val* by underscoring text. The coercions between trivial text and its meaning will be omitted henceforth.

By the reduction above, any function that satisfies *FAC's* defining equation must map *2* to *2!*. On the other hand, the expression "*FAC(-1)*" cannot be reduced to a value using the rules of Definition 2.2-2; *val* $[\![FAC(-1)]\!]$ is undefined. The *solution* of a specification is taken to be the set of minimally defined functions that satisfy their definitions. Minimality insures uniqueness and makes the specification language unambiguous. A formal development of this subject can be found in Manna's text (1974). The solution to *FAC's* defining equation is the function *factorial:Int → Int*

$$factorial(n) = \begin{cases} n! \ if \ n \geq 0 \\ undefined \ otherwise \end{cases}$$

Since solutions are unique we need not distinguish function variable symbols from the functions they denote. The name *FAC* rather than the name *factorial* can serve to identify *FAC's* solution.

Although we have now made subliminal any distinction between symbols and their denotations, we did not institute a formal connection between notation and its meaning merely to discard it in the next paragraph. We shall return to the definition of *val* when we discuss the mechanical reduction of expressions in Chapters 4 and 5.

There is ample temptation to be clever when performing reductions. The third step of the reduction above produces the subterm *mpy(1, <u>FAC(dcr(1))</u>)*. It is intuitively reasonable to replace this term by *<u>FAC(0)</u>*, since *1* is a multiplicative identity. However, reducing *mpy(0, <u>FAC(-1)</u>)* to *0* is suspect, since one of the subterms is undefined. While such "optimizing" simplifications make sense in computer arithmetic, we shall prohibit them by requiring that simplification only be applied to *convergent* terms, that is, to terms that are known to simplify to values.

DEFINITION 2.2-3. An operation (or predicate) f is strict *if it is undefined whenever any of its arguments is undefined. Strict operations that also respect φ, so that f(..., φ ,...) = φ as long as no arguments are undefined, are said to be* completely strict.

It is always assumed that the operations and predicates are strict but not always completely strict. The assumption implies that text cannot be "thrown away" through simplification in a reduction. The conditional reduction rule is therefore crucial, since by it alone may divergent subexpressions be discarded.

2.3. Reasoning about Specifications

We shall mainly use induction to reason about specifications. The methods used most are *structural induction* and *subgoal induction,* illustrated below. The examples in this section are based on recursion schemes or on recursion equations over **Int**. Infix notation for the arithmetic operations and predicates is used in order to make the examples easier to follow. Later, we shall revert to prefix notation.

2.3.1. Structural Induction. *Structural Induction* is the familiar technique for proving a proposition over an inductively defined set. To show a proposition P is true for all elements of a set S, one gives a proof "template" for a parameterized version of P, $P(s)$. In a *base* step, $P(s)$ is proven directly for the minimal elements in S. In an *induction* step, the assumption of $P(s)$ is shown to imply $P(s')$ where s' is any "next" element of S. For example,

PROPOSITION 2.3-1. Let G be defined as follows over Int:

$$G(x, y, z) \Longleftarrow (x = 0) \rightarrow y, G(x-1, z, y+z).$$

Then for all $a \geq 0$ and for all b and c, $G(a+2, b, c) = G(a, b, c) + G(a+1, b, c)$.
PROOF: By induction on *Int.* Let $P(k)$ be

"*For all b and c, $G(k+2, b, c) = G(k, b, c) + G(k+1, b, c)$.*"

Base step [$P(0)$].

$$
\begin{array}{ll}
G(2, b, c) = G(1, c, b+c) & \Delta G \\
\quad = G(0, b+c, b+2c) & \Delta G \\
\quad = b+c & \Delta G \\
\quad = G(0, b, c) + G(0, c, b+c) & \nabla G, \textit{twice} \\
\quad = G(0, b, c) + G(1, b, c) & \nabla G
\end{array}
$$

Induction [$P(k) \supset P(k+1)$]. Assume $G(k+2, b, c) = G(k, b, c)+G(k+1, b, c)$.

$$G(k+3, b, c) = G(k+2, c, b+c) \qquad \Delta G, \; k+3 \neq 0$$
$$= G(k, c, b+c) + G(k+1, c, b+c) \qquad \text{Induction Hypothesis, } P(k)$$
$$= G(k+1, b, c) + G(k+2, b, c) \qquad \nabla G, \; k+i \neq 0$$

□

The following corollary to Proposition 2.3-1 is used later.

COROLLARY 2.3-2. Let FIB and G be defined by

$$FIB(x) \Leftarrow (x \leq 1) \to 1, \; FIB(x\text{-}2) + FIB(x\text{-}1)$$

$$G(x, y, z) \Leftarrow (x = 0) \to y, \; G(x\text{-}1, z, y+z).$$

Then for all a ≥ 0, FIB(a) = G(a, 1, 1).

PROOF: The proof is by induction on *Int* using induction hypothesis
"*If a $\leq$ k+1 then FIB(a) = G(a, 1, 1)*". The details are given in Appendix C. □

2.3.2. Subgoal Induction.

Subgoal induction is an induction over the "depth" of recursion. The proof style, introduced by Morris and Wegbreit (1977), is natural because it uses the specification text as a proof generator. Hence, it emphasizes the notion that in writing a specification, the designer is in fact formulating a proof. Assume that all defining expressions are in *branched conditional* format:

$$F(x_1, ..., x_n) \Leftarrow p_1 \to r_1, \; p_2 \to r_2, ..., \; p_m \to r_m$$

where the propositional expressions p_i are mutually exclusive and exhaustive, and each r_j is a recurrent term. An *input-output assertion* $\Psi_F (x_1, ..., x_n \, ; z)$ is associated with each function variable symbol, relating its arguments x_i to its result z. Each branch of F's defining equation generates a *verification condition* of the form[3] $P \, \& \, I \supset R$. P is the premise that the predicate for the branch is true. I is

---

[3]A third premise is sometimes needed, stating that F produces equal outputs on equal inputs. This condition is not used in any of the proofs below.

the inductive assumption that all serious functions used in the branch satisfy their input-output assertions. The conclusion R states that the input-output assertion is true on this branch.

PROPOSITION 2.3-3. Let E be defined by

$$E(x) \Longleftarrow (x = 0) \rightarrow x,$$
$$(x \neq 0) \rightarrow E(x-1) + 2x - 1.$$

Then for all x, $E(x) = x^2$.

PROOF: by subgoal induction on E. E's input-output assertion is

$$\Psi_E(x\,;\,z) = \text{``}z = x^2.\text{''}$$

E's defining equation generates two verification conditions

   i. $(x = 0) \supset \Psi_E(x\,;\,x)$

   ii. $[(x \neq 0) \, \& \, \Psi_E(x-1\,;\,z)] \supset \Psi_E(x\,;\,z + 2x - 1)$

For verification condition *(i)*,

$x = 0$	*premise P*
$0 = 0^2$	*arithmetic fact*
$x = x^2$	*substitution of equals*

For verification condition *(ii)*,

$x \neq 0$	*premise P*
$z = (x-1)^2$	*premise I, that is $\Psi_E(x-1\,;\,z)$*
$z = x^2 - 2x + 1$	*arithmetic*
$x^2 = z + 2x - 1$	*more arithmetic*

The last line is $\Psi_E(x\,;\,z + 2x - 1)$ with $z = E(x-1)$. That is, if $x \neq 0$ then $E(x) = E(x-1) + 2x - 1 = x^2$. Since the predicates are exhaustive, the two cases establish the desired result.

□

Note that the undefined function $G(x) \Leftarrow true \rightarrow G(x)$ satisfies any input-output assertion. The function E in the example above does not meet its input-output assertion if it is given a negative argument, since it diverges. Subgoal induction is a *partial correctness* method; the functions involved satisfy their input-ouput assertions whenever they are defined. To show total correctness a separate termination proof may be given, or a well-founded measure may be included in the input-output assertion.

Subgoal induction is often used when not enough is known about the underlying type to support a structural induction. Hence, it useful for reasoning about recursion schemes, as the following proposition illustrates.

PROPOSITION 2.3-4. Let g be a commutative, associative, 2-place operation (i.e. for all x and y, $g(x, y) = g(y, x)$ and $g(x, g(y, z)) = g(g(x, y), z)$). Let G be defined by

$$G(x, y) \Leftarrow p(x) \rightarrow y, \neg p(x) \rightarrow G(h(x), g(x, y)).$$

Then for all a, b, and c, $G(a, g(b, c)) = g(b, G(a, c))$.

PROOF: by subgoal induction on G.
Case 1. If $p(a)$ is true, then by G's defining equation,
$$G(a, g(b, c)) = g(b, c) = g(b, G(a, c)).$$

Case 2. Assume that $p(a)$ is false, and by induction that for all b' and c',
$G(h(a), g(b', c')) = g(b', G(h(a), c'))$

$$
\begin{array}{ll}
G(a, g(b, c)) = G(h(a), g(a, g(b, c))) & \quad \Delta G, \neg p(a) \\
\qquad\qquad = G(h(a), g(b, g(a, c))) & \quad g \text{ is commutative and associative} \\
\qquad\qquad = g(b, G(h(a), g(a, c))) & \quad I.H.; \ b' = b, \text{ and } c' = g(a, c) \\
\qquad\qquad = g(b, G(a, c)) & \quad \nabla G, \neg p(a)
\end{array}
$$

□

Proposition 2.3-4 also has a useful corollary:

28

COROLLARY 2.3-5. Let FAC and G be defined by

$$FAC(x) \Leftarrow (x=0) \rightarrow 1, \ x * FAC(x\text{-}1).$$

$$G(x, \ y) \Leftarrow (x=0) \rightarrow y, \ G(x\text{-}1, \ x * y).$$

Then for all a $\geq$ 0, FAC(a) = G(a, 1).

PROOF: by structural induction on *Int.* See Appendix C.

□

2.4. Transformations on Recursion Equations

This section presents the central issue of this chapter: the translation of specifications from one form to another. For our purposes, the goal is to find a target specification that is in iterative form. Iterative form is of interest in general because of its correspondence to sequential control algorithms (*i.e. programs*) (McCarthy, 1963a; Patterson and Hewitt, 1970; Manna, 1974; Greibach, 1975). Since digital circuits are also sequential in nature, many of the results of research in compilation of recursion equations are also of use in the synthesis of circuits. The compilation problem has been studied widely; Cohen gives a survey of relevant papers in his dissertation (1980).

We embark on a series of extensions to the specification language that make it possible to find iterative "versions" of certain recursion structures. The first extensions are utterly reasonable; they express ways that basic components might be physically combined. Later extensions force us to make assumptions about the computational power of the underlying type; they yield iterative versions through constructions that implement recursion.

With modest extensions to the notation any iterative specification can be transformed to an instance of the "universal iterative scheme"

$$F(x) \Leftarrow p(x) \rightarrow f(x), \ F(g(x)). \qquad (U_I)$$

The initial connection between specifications and circuit descriptions is made on the basis of U_I. A collection of results, reviewed below, shows that any linear specification has an iterative version, although it may not compute in the same way as the original[4].

The simple extensions are not enough for more complex cases. Non-linear specifications exist for which no iterative version can be found, unless further assumptions are made about the underlying type. Corollary 2.3-5 is an example. It gives an efficient iterative version of the *factorial* specification, but the transformation depends on the algebraic properties of multiplication. As stronger assumptions are made about what can be computed by the underlying type, larger classes of specifications become transformable. It is not the purpose here to explore these relationships in detail. We shall simply stipulate that transformation is a "creative" design, meaning that intelligent guidance is permitted. Deriving a specification in a particular structural class is one heuristic for guiding the engineer.

2.4.1. Grammatical Transformations.

We shall refer to any "preprocessing" translation of a specification as a *grammatical transformation*. Such transformations are used to place specifications into a normal form in order to apply a general construction. Such translations exploit Definition 2.2-2 by symbolically folding or unfolding defining expressions. New definitions may be introduced into the system so that existing definitions can be folded into a simpler form. We shall see examples of this process in later derivations (*e.g.* in Section 2.4.4).

Branched Conditional Format.

In Section 2.3, recursion equations were assumed to have the form

$$F(x_1, ..., x_n) \Leftarrow p_1 \rightarrow r_1, \ p_2 \rightarrow r_2, ..., \ p_m \rightarrow r_m$$

Translation to this form would introduce additional function calls to replace r_i if it were a not a term, and modify the propositional expressions to make them mutually exclusive.

---

[4]This is a fuzzy qualification at best, since no measure of performance has been assigned to the specification language. Strong (1971) develops a formalization of *operational translatability* to address this issue.

Balanced Form. A specification is *balanced* if each defining expression in the system is a recurrent term, or a conditional whose alternatives are either both trivial or both serious. Extraneous function definitions can be used to balance alternatives. If the initial system is linear (iterative), a linear (iterative) balanced version can always be found (Greibach, 1975, pp 7–12).

Argument Padding. In constructions that follow it will be necessary to alter specifications so that each defining equation uses the same formal parameter list. The translation involves changing identifier names in a consistent fashion, and possibly adding unused formal parameters. By convention, the don't-care value ϕ is supplied as an argument when the corresponding formal parameter is padding.

2.4.2. Distributivity of the Conditional and Multiplexors.

The conditional construct distributes through application. For instance, the expressions $p \rightarrow f(r, s), f(t, u)$ and $f([p \rightarrow r,t], [p \rightarrow s,u])$ are equivalent, even if f is replaced by a function variable symbol[5]. While the non-strictness of the conditional is crucial to expression valuation, it is desirable to introduce a selective operation to replace conditional expressions when strictness isn't an issue. A *multiplexor* is a strict version of the conditional expression.

DEFINITION 2.4-1. Let p be a propositional expression. The operation mux *is defined as follows*

$$mux(p, b, c) = \begin{cases} b \text{ if } p \text{ is true} \\ \text{undefined if } p, b, \text{ or } c \text{ is undefined} \\ c \text{ if } p \text{ is false.} \end{cases}$$

Giving *mux* the status of an operation raises several technical problems. One of its operands is a propositional expression, which must now be admitted as a possible term. This forces truth values into the underlying type, and the remaining operations must be extended to handle them. We may assume either

---

[5]provided the function depends on one of its parameters. Consider $F(x) \Leftarrow c$. If $p \rightarrow F(a), F(b) = F([p \rightarrow a,b])$, then it reduces to c, regardless of p. Definition 2.2-2 suggests that conditionals should be undefined if their propositional expressions are.

that the underlying type "admits selection", perhaps through an encoding of truth values in the carrier, or that the valuation function has been patched with a special case for multiplexors. In any event, the issue is not crucial because multiplexors are only used here to replace conditionals.

Transforming "$p \rightarrow r, s$" to "$mux(p, r, s)$" is tantamount to an assertion that r and s both converge. For example, replacing

$$F(x) \Longleftarrow p(x) \rightarrow f(x), f(F(x)).$$

by

$$F(x) \Longleftarrow f(mux(p(x), x, F(x))).$$

is invalid because in the second form, the defining expression always diverges, whereas the first does not. The following criteria are sufficient to guarantee that replacement by multiplexors is harmless:

1. r and s are trivial expressions.
2. The surrounding specification is linear.

The conditions insure that r and s will be ground terms in any reduction, and will therefore always converge. If the surrounding specification is linear, p must be trivial by Definition 2.1-3. By condition (1) r and s contain no serious subexpressions, and divergence cannot be introduced through unfolding the recursion equation in which they occur. Condition (2) implies that no prior substitution has introduced a serious expression.

2.4.3. Combined Operations.

The notation is now extended to permit groups of operations to be expressed as a single *combined* operation.

Constants. For each constant c introduce a constant-operation with symbol K^c.

$$K^c(z) = c.$$

Identifiers. For each coordinate of a vector, introduce a *projector*, π_i .

$$\pi_i(x_1, x_2, ..., x_n) = x_i.$$

Serial combination. Operator composition is expressed by juxtaposition.

$$f \circ g(z) = f(g(z)).$$

Parallel combination. A sequence of operations enclosed in angle brackets denotes "broadcast" of the argument.

$$<f_1 ,..., f_n>(z) = (f_1(z) ,..., f_n(z)).$$

These extensions make sense in terms of circuitry. As their names indicate, parallel and serial combination suggest ways that components are physically wired together. Projection is a "tie into a bus". A constant-operation corresponds to a fixed source.

The goal is to rewrite any terminal term over the identifiers $x_1 ,..., x_n$ as something of the form $\gamma(x_1 ,..., x_n)$, where γ is a combined operation. In the process, individual identifiers are replaced by their coordinate addresses in an argument vector. The combined term may be written simply as $\gamma(z)$, where the identifier z stands for the argument vector.

PROPOSITION 2.4-2. Define a translator τ, taking terminal terms to combined operations, as follows

$$\tau [\![\, c \,]\!] = K^c$$

$$\tau [\![\, x_i \,]\!] = \pi_i$$

$$\tau [\![\, f \lfloor\, t_1 \llcorner \cdots \llcorner t_n \,\rfloor \,]\!] = f < \tau [\![\, t_1 \,]\!] \cdots \tau [\![\, t_1 \,]\!] >$$

For any terminal term t over the identifiers $x_1 ,..., x_n$

$$val [\![\, \tau [\![\, t \,]\!](a_1 ,..., a_n) \,]\!] = val [\![\, t \begin{bmatrix} a_1 ,..., & a_n \\ x_1 ,..., & x_n \end{bmatrix} \,]\!]$$

PROOF: The proof is a straightforward structural induction on L_T but requires a formal definition of substitution. Several similar proofs may be found in Wand's text (1980).

□

Combined operations will be introduced exclusively by the translator . The underlying type is not necessarily closed under arbitrary combinations, for if it were, they could be used to build data structures. As with multiplexors, the use of combinations is limited to cases where they can be dealt with syntactically by an enhanced valuation function. They serve simply as "macros".

Notice that the term $f(c)$ translates to $f<K^c>$. But by the definitions above of serial and parallel combination,

$$f<K^c>(z) = f(K^c(z)) = f{\circ}K^c(z)$$

Although the translator encloses all argument lists, even those of length one, in a parallel combination, we shall suppress the brackets in the case of 1-place function combinations for the sake of brevity[6]. Thus $f<K^c>$ is written $f{\circ}K^c$.

2.4.4. Universal Schemes.

Specifications can be classified by a collection of representative schemes to which they can be transformed. Using grammatical transformations, multiplexors, and combined operations, there is a construction by which any iterative specification can be transformed to an instance of the scheme

$$F(x) \Longleftarrow p(x) \to f(x),\ F(g(x)). \qquad (U_I)$$

The construction is straightforward, and is roughly the same as Cooper's version (1967) of the folk theorem: "Every looping structure can be transformed to a single while-loop" (Harel, 1980). However, it is carried out in a functional notation. We will make do with a small example, itself a generalization that shows how to construct iterative versions of certain linear specifications. Consider the recursion scheme:

$$L(x) \Longleftarrow p(x) \to f(x),\ h(L(g(x))$$

and the iterative system

$$R_0 \quad \boxed{\begin{array}{l} G(x,\ y) \Longleftarrow p(x) \to H(y,\ f(x)),\ G(g(x),\ y). \\ H(x,\ y) \Longleftarrow p(x) \to x,\ H(g(x),\ h(y)). \end{array}}$$

L returns $h^n f g^n(x)$, where the superscript denotes n-fold composition, and n is the number of times g must be applied to x in order to make p true. Intuitively, G computes $f g^n(x)$ and passes it to H, along with the initial value of x. H uses p to

---

[6]A quite elegant approach to programming results from the algebra of combinations in which this transformation is an elementary rule (See Backus, 1978, 1981). The use of combined operations is transitory in this presentation; it lasts until Section 3.4.

recompute n, and applies h that many times. It is not difficult to show that

PROPOSITION 2.4-3. For all a, L(a) = G(a, a).

PROOF: (Appendix C).

We shall now construct an instance of U_I from specification R_0. The construction requires the initial system to be in balanced form (Sec. 2.4.1). To balance our example, we need only replace the x in H's defining equation with a dummy function call. I's defining equation is padded to make its formal parameter list conform to the others.

$$R_1 \quad \boxed{\begin{array}{l} G(x,\ y) \Longleftarrow p(x) \to H(y,\ f(x)),\ G(g(x),\ y). \\ H(x,\ y) \Longleftarrow p(x) \to I(x,\ \phi),\ H(g(x),\ h(y)). \\ I(x,\ y) \Longleftarrow x. \end{array}}$$

The next step introduces a new parameter to record which function is "in control", and rewrites the system as a single recursion equation. It must be assumed that the encoding can be represented in the underlying type. Let *control token* w range over { **G**, **H**, **I** } and let the predicate *at?* be a test for one of these values. Transform R_1 into a single defining equation for function F:

$$R_2 \quad \boxed{\begin{array}{l} F(w,\ x,\ y) \Longleftarrow at?(w,\ \mathbf{I}) \to x, \\ \qquad\qquad at?(w,\ \mathbf{G}) \to [\ p(x) \to F(\mathbf{H},\ y,\ f(x)),\ F(\mathbf{G},\ g(x),\ y)\ ], \\ \qquad\qquad\qquad [\ p(x) \to F(\mathbf{I},\ x,\ \phi),\ F(\mathbf{H},\ g(x),\ h(y))\ ]. \end{array}}$$

The propositions distribute. We first push p inside the call to F; since the scheme is linear, multiplexors can be used for selection.

$$R_3 \quad \begin{array}{l} F(w,\ x,\ y) \Leftarrow \\ \quad at?(w,\ \mathbf{I}) \to x, \\ \quad at?(w,\ \mathbf{G}) \to F(mux(p(x),\ \mathbf{H},\ \mathbf{G}),\ mux(p(x),\ y,\ g(x)),\ mux(p(x),\ f(x),\ y)), \\ \qquad\qquad F(mux(p(x),\ \mathbf{I},\ \mathbf{H}),\ \ mux(p(x),\ x,\ g(x)),\ mux(p(x),\ \phi,\ h(y))). \end{array}$$

Distribution of *at?* yields

$$R_4 \quad \begin{array}{l} F(w,\ x,\ y) \Leftarrow at?(w,\ \mathbf{I}) \to x, \\ \qquad F(mux(at?(w,\ \mathbf{G}),\ mux(p(x),\ \mathbf{H},\ \mathbf{G}),\ \ mux(p(x),\ \mathbf{I},\ \mathbf{H})), \\ \qquad\quad mux(at?(w,\ \mathbf{G}),\ mux(p(x),\ y,\ g(x)),\ mux(p(x),\ x,\ g(x))), \\ \qquad\quad mux(at?(w,\ \mathbf{G}),\ mux(p(x),\ f(x),\ y),\ mux(p(x),\ \phi,\ h(y))). \end{array}$$

The operations of R_4 are combined to get the desired instance of U_l. Let

$$p' = at? <\pi_1\ K^I>$$

$$f' = \pi_2$$

$$g' = <\ mux<at?<\pi_1\ K^G>\ \ mux<p\circ\pi_2\ K^H\ K^G>\ \ mux<p\circ\pi_2\ K^I\ K^H>>$$

$$\qquad mux<at?<\pi_1\ K^G>\ \ mux<p\circ\pi_2\ \pi_3\ g\circ\pi_2>\ \ mux<p\circ\pi_2\ \pi_2\ g\circ\pi_2>>$$

$$\qquad mux<at?<\pi_1\ K^G>\ \ mux<p\circ\pi_2\ f\circ\pi_2\ \pi_3>\ \ mux<p\circ\pi_2\ K^\phi\ h\circ\pi_3>>>$$

Using these combined operator symbols we arrive at the desired instance of U_l:

$$R_5 \quad F(z) \Leftarrow p'\,(z) \to f'\,(z),\ F(g'\,(z)).$$

The construction preserves the meaning of the initial specification. It can be shown by subgoal induction on F that for all a and b

$$F(\mathbf{G},\ a,\ b) = G(a,\ b),$$
$$F(\mathbf{H},\ a,\ b) = H(a,\ b),\ and$$
$$F(\,\mathbf{I},\ a,\ b) = I(a,\ b).$$

Hence by Proposition 2.4-3, $F(\mathbf{G},\ a,\ a) = L(a)$, where L is defined by the linear equation we began with. The construction can clearly be generalized to arbitrary iterative systems, and a similar construction yields a universal linear scheme.

THEOREM 2.4-4. *If multiplexors and combined operations are allowed then any* *iterative specification can be transformed to an instance of the scheme* U_I:

$$F(x) \Leftarrow p(x) \to f(x),\ F(g(x)).$$

Any linear specification can be transformed to an instance of the scheme U_L:

$$F(x) \Leftarrow p(x) \to f(x),\ h(x,\ F(g(x))).$$

PROOF: Each scheme is a special case of a construction presented by Cohen (1980, pp. 639-643), who cites Chandra as the originator of U_L (Chandra, 1972).

$\square$

Patterson and Hewitt (1972) also note the universality of U_L when they present a flowchart schema equivalent to any linear specification. The following theorem restates their result as an assertion about transformability to iterative form.

THEOREM 2.4-5. *Let F be defined by* U_L, *and consider the specification.*

$$G(u,\ v,\ x,\ y,\ z) \Leftarrow p(x) \to L(u,\ ,\ u,\ ,\ fx),$$
$$G(u,\ ,\ gx,\ ,\ ).$$

$$L(u,\ v,\ x,\ y,\ z) \Leftarrow p(x) \to z,\ M(u,\ gx,\ gx,\ u,\ z).$$

$$M(u,\ v,\ x,\ y,\ z) \Leftarrow p(x) \to L(u,\ ,\ v,\ ,\ h(y,\ z)),$$
$$M(u,\ v,\ gx,\ gy,\ z).$$

For all a, $F(a) = G(a,\ \phi,\ a,\ \phi,\ \phi).$
PROOF: (Appendix C).

The extensions allowed so far are not powerful enough to yield iterative versions of arbitrary specifications. The following well known example is due to Patterson and Hewitt (1972):

THEOREM 2.4-6. *If multiplexors and structural combination are all that is* *allowed, there is no general transformation that yields an iterative version of*

$$F(x) \Leftarrow p(x) \to f(x),\ h(\ F(g_1(x)),\ F(g_2(x))\ ).$$

DISCUSSION: The usual statement of the theorem is that the scheme is not "flowchartable". Its proof depends on formalizations we have not introduced and so it is omitted. The strategy is to show that the iterative version would need an unbounded number of identifiers to produce the right value in an arbitrary underlying type. For details see (Patterson and Hewitt, 1972), (Manna, 1974), or (Greibach, 1975).

□

2.4.5. Synthesis of Iterative Form

Specifications in iterative form correspond with the notion of sequential control associated with flowcharts; a program statement is a function on the program's state. We have assembled enough notation to permit any linear specification to be translated to iterative form and hence to an instance of the scheme U_I. So far, we have made simple stipulations about the computational qualities of the underlying type. It must admit selection and certain forms of combination. It must be robust enough to represent a finite number of control tokens and have a test for equality. The review included the negative result that not all specifications have iterative versions.

On the basis of recursive structure alone, it is not decidable whether a non-linear specification has an iterative equivalent (see for instance Greibach, 1975, Theorem 7.9). However, in the course of our discussions we have managed to find iterative versions of all of our example specifications. Corollary 2.3-5 shows that by introducing an "accumulator", the *factorial* specification

$$FAC(x) \Leftarrow zero?(x) \to 1,\ mpy(x,\ FAC(dcr(x))).$$

has iterative version

$$G(x,\ y) \Leftarrow zero?(x) \to y,\ G(dcr(x),\ mpy(x,\ y)).$$

This version is intuitively better than the construction of Theorem 2.4-5 because it is faster; but its validity depends on the algebraic properties of multiplication.

Corollary 2.3-3 shows that the *Fibonacci* specification:

$$FIB(x) \Leftarrow lt?(x,\ 2) \to 1,\ add(\ FIB(dcr(dcr(x))),\ FIB(dcr(x))\ ).$$

Has iterative version

$$G(x,\ y,\ z) \Leftarrow zero?(x) \to y,\ G(dcr(x),\ z,\ add(y,\ z)).$$

Hence, not all instances of the troublesome non-linear scheme of Theorem 2.4-6 resist translation.

Cohen (1980) reviews efforts to address the translation problem. The work generally follows two lines, both of which are forms of synthesis. Darlington and Burstall (1977) describe "an inference system in which the sentences are recursion equations" where human guidance adds information that makes transformation succeed. A specification is transformed algebraically by folding, unfolding, and the application of previously established transformation rules, until an improved specification emerges. The other approach is to assume that explicit operations exist or can be implemented in the underlying type, in effect supposing it can be used to implement certain recursion patterns. More powerful operations permit wider specification classes to be "linearized."

Having looked at the transformation-system approach let us now consider the recursion-implementing strategy. Suppose that the underlying type contains operations that are powerful enough to implement *stacks*. That is, assume that a value ϵ; completely strict combinations called *push*, *pop*, and *top*; and propositional combination *empty?* ; all exist that satisfy

$$empty?(\epsilon) = true \qquad empty(push(u, v)) = false$$
$$top(\epsilon) = \phi \qquad pop(\epsilon) = \phi$$
$$top(push(u, v)) = u \qquad pop(push(u, v)) = v$$

If these powerful operations are available, then general methods exist to linearize arbitrary specifications. The construction below, due to Wand and Friedman (1978), is used in Chapter 5. It introduces a "run time stack" and a new serious function to handle "return jumps". The specification is then repeatedly refined so that control is linearized.

CONSTRUCTION 2.4-7. (Wand and Friedman, 1978)

For simplicity, assume that in the initial specification all functions are defined over the same set of identifiers.

$$F_1(x_1, ..., x_n) \Leftarrow e_1.$$
$$\vdots$$
$$F_m(x_1, ..., x_n) \Leftarrow e_m.$$

Designate a set of *action* values, $\{a_1, ..., a_k\}$ where k will be determined by the time the transformation is complete, and rewrite each equation as

$$F_i\,(x_1,...,\,x_n,\,\sigma) \Longleftarrow R(e_i\,,\,\sigma).$$

The new parameter σ names the recursion stack. Add a new function variable symbol R, for "return", whose defining equation is constructed as we go along. Its general form will be

$$
\begin{aligned}
R(v,\,\sigma) \Longleftarrow\ & empty?(\sigma) \to v, \\
& eq?(top(\sigma),\,a_1) \to\ do\text{-}something\text{-}with\text{-}v\text{-}and\text{-}restore\text{-}\sigma, \\
& eq?(top(\sigma),\,a_2) \to\ do\text{-}something\text{-}with\text{-}v\text{-}and\text{-}restore\text{-}\sigma, \\
& \qquad\qquad\vdots \\
& eq?(top(\sigma),\,a_k) \to\ do\text{-}something\text{-}with\text{-}v\text{-}and\text{-}restore\text{-}\sigma.
\end{aligned}
$$

Arbitrarily select a serious expression of the form $R(e,\,\sigma)$ and transform the system as follows

1. (Tail-recursive call) If e is of the form $F_i(t_1,...,\,t_n)$, and each t_j is trivial, change $R(e,\,\sigma)$ to

   $$F_i(t_1,...,\,t_n,\,\sigma).$$

2. (Decision) If e is of the form $p(t_1,...,\,t_q) \to r,\,s$, and each t_j is trivial, change $R(e,\,\sigma)$ to

   $$p(t_1,...,\,t_q) \to R(r,\,\sigma),\,R(s,\,\sigma).$$

3. If e is not in any of the forms above, then find an expression e' over unused identifiers $y_1,...,\,y_m$; a serious expression r; and trivial expressions $t_2,...,\,t_m$; such that

   $$e = e'\begin{bmatrix} r,\ t_2,...,\ t_m \\ y_1,\ y_2,...,\ y_m \end{bmatrix}$$

   If e is a conditional, choose r from its propositional expression if possible. Obtain an unused action value a, and replace $R(e,\,\sigma)$ by

   $$R\big(r,\,push(a,\,push(t_2,...,\,push(t_m,\,\sigma)...)\,)\,\big)$$

   and add to R's defining equation the clause

   $$eq?(top(\sigma),\,a) \to R\Big(\,e'\begin{bmatrix} v,\ s_2,...,\ s_m \\ y_1,\ y_2,...,\ y_m \end{bmatrix},\,pop^m(\sigma)\,\Big)$$

   where s_i stands for the term $top(pop^i(\sigma))$.

In words, Step 3 says to pick a serious term to call recursively. Any trivial values needed on the return may be computed now and saved on the stack. By the time the transformation is complete the stack parameter will have been introduced to all serious calls, and the specification will be in iterative form.

EXAMPLE 2.4-8. If the Wand-Friedman construction is applied to the specification

$$F(x) \Leftarrow p(x) \to c, \ h(\ F(g_0(x)), \ F(g_1(x)) \ ).$$

one possible target specification is

$F(x, \sigma) \Leftarrow p(x) \to R(c, \sigma), \ F(\ g_0(x), \ push(0, \ push(g_1(x), \ \sigma))).$

$R(v, \sigma) \Leftarrow empty?(\sigma) \to v,$
$\qquad at?(top(\sigma), \ 0) \to F(\ top(pop(\sigma)), \ push(1, \ push(v, \ pop(pop(\sigma)))) \ ),$
$\qquad at?(top(\sigma), \ 1) \to R(\ h(top(pop(\sigma)), \ v), \ pop(pop(\sigma)) \ ).$

The derivation is shown in Appendix C.

This construction does not state how to choose which serious term to call. In the example, the strategy was to evaluate arguments left-to-right. The obvious criterion is to choose an expression that is known to be needed. Mycroft (1980) gives an algorithm that makes this determination under certain conditions. If the choice is wrong, the target specification may errantly diverge. Since partially correct target specifications are sometimes acceptable, we shall leave this choice to the designer's discretion.

2.5. The Scott-Strachey Notation

This section is a brief review of the "type free" notation of Scott and Strachey. It is a language defined over a universal type in which any reasonable (*i.e.* computable) type can be embedded. It extends the specification language to a most-abstract realm, and for that reason its use as a design language always entails choosing an appropriate type over which a more concrete specification can be made. The Scott-Strachey style has been used mainly to describe programming languages, and a rather rich notation has evolved from this application. Tennent (1976), Gordon (1979), and Scott (1977, 1982) each give a casual introduction to the notation and its use. The standard text on the subject is by Stoy (1977). The two volume work of Milne and Strachey (1976) is a comprehensive example of the use of the theory to describe a programming language.

For our purposes Scott's is a theory of data types, which he calls *domains*. One may think of a domain as a set of descriptions, or answers that might be printed by a program. Some descriptions are better than others, in the sense that they are more complete; some are incomparable because they do not describe the same thing.

A *domain* then, is a set D with a reflexive, transitive relation called *approximation* and expressed by the symbol "$\sqsubseteq$". Membership in D is expressed by the symbol "$\in$". D must satisfy certain axioms with respect to $\sqsubseteq$. It must contain a minimal[7] (or empty, or divergent) description, denoted by $\perp_D$, that approximates every other description. That is,

$$\textit{for all } d \in D, \ \perp \sqsubseteq d.$$

Intuitively, any sequence of successively better elements in D must converge to a limit that is also in D. Operations on domains are required to be *continuous*, that is, to preserve limits[8].

2.5.1. Flat Domains.
A basic, or *flat*, domain meets the minimal requirements:

$$x \sqsubseteq y \textit{ iff } x = \perp \textit{ or } x = y.$$

Examples:

Truth values — $Bool = \{ \perp, \textit{tt}, \textit{ff} \}$.

Integers — $Int = \{ \perp \} \cup \{..., -2, -1, 0, 1, 2, ...\}$

---

[7]A maximal (or overdefined, or contradictory) description may also be assumed. It does not enter into any of the discussions in later chapters, and so is ignored in this review.

[8]Continuous functions preserve limits over a wider class of sets than those that are monotone. Monotonicity can be generalized to "directedness" (Stoy, 1977). The real concern is not with individual descriptions, but with *neighborhoods:* collections of approximations to the same ideal. Scott has recently rephrased his formal presentation in these terms (Scott, 1982).

Numerals — $Nml = \{\perp\} \cup$ *strings over* $\{\underline{0}, \underline{1},..., \underline{9}\}$

Identifiers — $Ide = \{\perp, \underline{u}, \underline{v}, \underline{w}, \underline{x}, \underline{y}, \underline{z} ...\}$.

The conventional ordering on these sets (*e.g.* $\leq$ on the set of integers) is not the domain ordering: a program that is supposed to print '5' may diverge and produce no description, or it may print '5', but if it prints '4' then it is not an approximation. *Bool* and *Int* are *semantic* domains corresponding to the carriers of our underlying types. *Nml* and *Ide* are *syntactic* domains wherein we have defined our specification languages. We shall see later that the distinction is subjective.

2.5.2. Non-flat Domains

Complex domains are built by combining domains in various standard ways. Given domains A and B, the following domain constructors are needed:

A $\times$ B   The (coalesced) *product* domain is the set $\{(a, b) \mid a \in A \text{ and } b \in B\}$ with ordering $(a, b) \sqsubseteq_{A \times B} (a', b')$ *iff* $a \sqsubseteq a'$ *and* $b \sqsubseteq b'$.

$\perp_{A \times B} = (\perp_A, \perp_B)$ serves consistently as the minimal element.

A + B   The (separated) *sum* domain is the set $A \cup B \cup \{\perp_{A+B}\}$ with approximation ordering $x \sqsubseteq_{A+B} y$ *iff* $x = \perp_{A+B}$ *or* $x \sqsubseteq_A y$ *or* $x \sqsubseteq_B y$.

A $\rightarrow$ B   The *function* domain is the set of continuous functions from A to B, with approximation ordering $f \sqsubseteq_{A \rightarrow B} g$ *iff* $x \in A$ *implies* $f(x) \sqsubseteq_B g(x)$.

A^n The *n-ary product* domain is a generalization of the product domain construction to *n*-tuples, for a given *n*.

Let *e* be an expression, possibly including the identifier *x*, and suppose that whenever some element $a \in A$ is substituted for *x* in *e*, the result is a unique element of *B*. Hence, substitution induces a function from A to B. The *abstraction of e by x*, written "$\lambda x.e$", denotes the function just described. If *e* is suitably expressed, then this function is continuous; that is $(\lambda x.e) \in A \rightarrow B$. Applicative expressions like *f(x)*, abstractions themselves, and conditional expressions

$$p \rightarrow a, b \begin{cases} a \text{ if } p = tt \\ b \text{ if } p = ff \\ \perp \text{ if } p = \perp \end{cases}$$

are all suitable for abstraction. The nesting of λ-expressions gives rise to identifier conflicts, and so the substitution rules must be refined to replace only *free* identifiers—those that are not in the scope of an interior λ-expression.

There are several conventions for abbreviating this simple but verbose language. Since parentheses serve only to state the scope of an expression, they are often suppressed. One may write *"fa"* rather than *"f(a)"*. Application associates to the left, so that *"fga"* means *"(fg)a"*. Note that this differs from our convention for serial combination. The function-domain constructor associates to the right to be consistent with this convention. That is, $A \rightarrow B \rightarrow C = A \rightarrow (B \rightarrow C)$. When the context allows it, membership in a function-domain is expressed with a colon rather than the membership symbol " $\in$ ". Thus, *"f:A→B"* mimics the mathematical notation for saying *"f"* is a partial function from set A to set B."

The scope of a λ-expression extends as far to the right as possible, generally to the end of a line, or to the first unbalanced parenthesis. *"λuv . e"* is abbreviated to *"λu.λv.e"*; *"F(x) ⇐ e."* is another way of writing *"F = λx.e"*; and *"e where x = t"* means *"(λx.e)(t)"*.

2.5.3. Domain Operations.

For flat domains, continuous versions of basic operations may be assumed. There are standard operations to go with complex domain constructions. These operations are expressed with special notation.

Products. Let $D = A \times B$, $a \in A$, and $b \in B$. There is a pairing function $\langle *, * \rangle \in A \rightarrow B \rightarrow D$, and there are projectors $*{\downarrow}0 \in D \rightarrow A$ and $*{\downarrow}1 \in D \rightarrow A$ that satisfy[9]

$$\langle a, b \rangle {\downarrow} 0 = a \qquad \langle a, b \rangle {\downarrow} 1 = b$$

This notation may be extended to n-ary products.

Sums. Let $D = A + B$ and let $a' \in D$ and $b' \in D$ be elements that came from $a \in A$ and $b \in B$ respectively. Operations $*isA \in D \rightarrow Bool$ (*inspection*), $*inD \in A \rightarrow D$ (*injection*), and $*asA \in D \rightarrow A$ (*restriction*) exist that satisfy

$$a' isA = tt \qquad ainD = a' \qquad a' asA = a$$
$$b' isA = ff \qquad \qquad b' asA = \perp_A$$

The corresponding operations *isB*, *asB*, and *inB* exist for the summand B.

---

[9]Usually, $*{\downarrow}1$ and $*{\downarrow}2$ are used instead.

2.5.4. Functionals.

Functions qualify as data types, and there are a number of higher level functions-on-functions, or *functionals*, that are useful. Among these are the structural combinators (*cf.* Sec. 2.4.3):

$$K^c = \lambda z.c$$
$$\pi_1 = (\lambda z.z\!\downarrow\!0), \ \pi_2 = (\lambda z.z\!\downarrow\!1), \ etc.$$
$$f \circ g = \lambda fg.(\lambda z.f(gz))$$
$$<f_1 \ldots f_n> = \lambda f_1 \ldots f_n.(\lambda z.\langle f_1 z, \ldots, f_n z \rangle)$$

The following standard functionals are used later.

apply:$(A \rightarrow B) \rightarrow A \rightarrow B$ takes a function and an argument and returns the correct answer for that function on that argument.

$$apply = \lambda fa.fa$$

curry:$((A \times B) \rightarrow C) \rightarrow (A \rightarrow B \rightarrow C)$ takes a 2-place function and returns a 1-place function that must be applied twice to get the desired value.

$$curry = \lambda f.\lambda u.\lambda v.f \langle u, \ v \rangle.$$

For example, if *add : (Int $\times$ Int) $\rightarrow$ Int* is the 2-place addition function, then *(curryadd)(2)* returns the function that adds two to its argument, and *(curryadd)(2)(2) = add$\langle 2, 2 \rangle$*. There is an inverse to *curry* that "unwraps" argument tuples.

$$uncurry = \lambda f.(\lambda x.f(x\!\downarrow\!0)(x\!\downarrow\!1))$$

Uncurry is expressed implicitly by enclosing formal parameters in square brackets, "$\lambda[u, v].e$".

2.5.5. Recursion.

If *f:D$\rightarrow$D*, then $d \in D$ is called a *fixed point* of *f* if $d = f(d)$. The function *fix:$(D \rightarrow D) \rightarrow D$* returns the minimal fixed point. That is, *(fix f) = f(fix f)* and for all fixed points *d* of *f*, *fix(f)* $\sqsubseteq$ *d*. *Fix* is continuous and expressible in λ-notation[10].

If D is itself a function domain, then *fix* yields the solution prescribed by Definition 2.2-2. For example, let $D = Int \rightarrow Int$ and take *f* to be

---

[10]One version is the *combinator* $Y = \lambda f.(\lambda x.f(xx))(\lambda x.f(xx))$. See (Stoy, 1977).

$$\lambda\xi.(\ \lambda n.(n{=}0) \rightarrow 1,\ n*(\xi(n{-}1))\ ).$$

Note that f has the required functionality, $D \rightarrow D$. If $FAC = fix(f)$ then $FAC = f(FAC)$. That is,

$$FAC = (\ \lambda\xi.(\ \lambda n.(n=0) \rightarrow 1,\ n*(\xi(n{-}1)))\ )(FAC)$$
$$= \lambda n.(n=0) \rightarrow 1,\ n*(FAC(n{-}1))$$

by substitution. Since *fix* gives the minimal solution we are justified in writing the equation above as

$$FAC(n) \Longleftarrow (n{=}0) \rightarrow 1,\ n*(FAC(n{-}1)).$$

On the other hand, this discussion shows that we can avoid self-reference in our specifications by using *fix* in writing

$$FAC = fix(\ \lambda\xi.[\ \lambda n.(n{=}0) \rightarrow 1,\ n*(\xi(n{-}1))\ ]\ )$$

2.5.6. Reflexivity. Fixed points can be defined over any domain, and *fix* is also used to define self-referential, or *reflexive* data types. For example, the domain of "s-expressions":

$$Sexp = Atom + (Sexp \times Sexp)$$

describes LISP's data space (McCarthy, 1960); an s-expression is either atomic or consists of a pair of s-expressions.

2.6. Other Issues

The additional power of the Scott-Strachey notation, the facilities to describe data structures and to manipulate functions, make it possible to attack aspects that are difficult to address in the more concrete typed language. Issues such as the specification of meaning—the original motive of Scott's and Strachey's work—and the formalization of "control" yield quite gracefully to the calculus. These topics and a few others are reviewed in this section, partly to exercise the rather extensive notation that has been introduced so far. Each of the issues discussed here arises later in the investigation.

2.6.1. Specifying the Specification Language. We give a brief example to demonstrate the use of Scott-Strachey notation to describe semantics. Consider the language of terminal terms defined in Section 2.2 (Definition 2.1-2). Suppose for simplicity that all operations take exactly two operands. The syntax of L_T can be defined as a reflexive data type

$$Term = Nml + Ide + Apl$$
$$Apl = Opr \times Term \times Term$$

Terms are built from atoms in the flat domains of numerals and identifiers and a collection of operation symbols. A domain *Apl* of applicative terms is recursively constructed by pairing an operation symbol with two operand-terms. Numerals denote integers and operators denote 2-place functions on integers. These meanings are also domains, namely *Int* and $Opn = (Int \times Int) \to Int;$ let them be given by semantic functions $\mathbb{N}:Nml \to Int$ and $\mathbb{K}:Opr \to Opn$. A mapping from identifiers to their meanings is also needed, and will be in the domain of environments $Env = Ide \to Nml$.

We are now ready to define a semantic function $\mathbb{T}: Term \to Env \to Int$ that specifies the meaning of a term:

$$\mathbb{T} = \lambda t\rho.(tisNml) \to \mathbb{N}(tasNml),$$
$$(tisIde) \to \mathbb{N}(\rho(tasIde)),\ help(tasApl)\rho.$$

$$help = \lambda a\rho.(\mathbb{K}(a{\downarrow}0)) \langle \mathbb{T}(a{\downarrow}1)\rho\ ,\ \mathbb{T}(a{\downarrow}2)\rho \rangle.$$

The auxiliary function *help* simply makes the definition easier to read. *Help* could be eliminated by expanding its definition in the equation for $\mathbb{T}$; and so it serves as a "macro". Expressions like *help*, which have no free variables and can therefore always be eliminated by substitution, are called *combinators*.

Additional abbreviations make these definitions easier to read. A Backus-Naur style is used to describe syntactic domains and to document concrete syntax. Valuation functions are written as a set of identities in the style of Definition 2.2-2. Elementary coercions are suppressed through the use of naming conventions. The revised definition is given in Figure 2.1. The figure gives a *standard* semantics for *Term*; it says nothing, for example, about the order of argument evaluation or error recovery. Stoy (1977) gives methods for addressing such issues, one of which is introduced in the following section. Language specifications in this style will be made in Chapters 4 and 5.

2.6.2. Specifying Control.

A *continuation* is a formalization of control in the domain *Value* → *Answer;* it takes a value produced in the present and states what is to be done with that value to produce an answer. One can linearize a non-linear specification by using continuations to describe a calling order.

Syntactic Domains

   Ide   *(i) identifiers*

   Nml   *(n) numerals*

   Opr   *(o) operators*

   Term ::= Nml | Ide | Opr (Term , Term)   *(t) terminal terms*

Semantic Domains

   $Opn = (Int \times Int) \rightarrow Int$ *operations*

   $Env = Ide \rightarrow Nml$ *(ρ) environments*

Valuations

   $\mathbb{K} : Opr \rightarrow Opn$

   $\mathbb{N} : Nml \rightarrow Int$

   $\mathbb{T} : Term \rightarrow Env \rightarrow Int$

$$\mathbb{T}[\![\, n \,]\!]\rho \; = \; \mathbb{N}(n)$$

$$\mathbb{T}[\![\, i \,]\!]\rho \; = \; \mathbb{N}(\rho i)$$

$$\mathbb{T}[\![\, o(\, t_1 , t_2 \,) \,]\!]\rho \; = \; (\mathbb{K}[\![\, o \,]\!])\langle\, \mathbb{T}[\![\, t_1 \,]\!]\rho \,,\, \mathbb{T}[\![\, t_2 \,]\!]\rho \,\rangle$$

Figure 2.1. A Standard Semantics for Terminal Terms.

For example, consider the following proposition:

PROPOSITION 2.6-1. Let F and G be defined as follows

$$F(x) \Longleftarrow p(x) \rightarrow c,\, h(\, F(g_0(x)),\, F(g_1(x))\,).$$

$$G(x, \kappa) \Longleftarrow p(x) \rightarrow \kappa(c),\, G(g_0(x),\, [\, \lambda u.\, G(g_1(x),\, [\, \lambda v.\kappa h(u,\, v)\,])\,])\,).$$

Then for all a and γ*,* $G(a, \gamma) = \gamma F(a)$*.*

PROOF: by subgoal induction on *F*. If *p(a)* is true then both sides reduce to
$\gamma(c)$. Otherwise,

$$G(a, \gamma) = G(g_0(a), [\lambda u.G(g_1(a), [\lambda v.\gamma h(u, v)])]) \qquad \Delta G$$

$$= [\lambda u.G(g_1(a), [\lambda v.\gamma h(u, v)])])(F(g_0(a))) \qquad I.H.$$

$$= G(g_1(a), [\lambda v.\gamma h(F(g_0(a)), v)]) \qquad substitution$$

$$= [\lambda v.\gamma h(F(g_0(a)), v)](F(g_1(a))) \qquad I.H.$$

$$= \gamma h(F(g_0(a)), F(g_1(a))) \qquad substitution$$

$$= \gamma F(a) \qquad \triangledown F$$

$$\square$$

A more palatable version of G results if we introduce names for its continuations.

$$G(x, \kappa) \Longleftarrow p(x) \rightarrow \kappa(c), \; G(g_0(x), DoG(g_1(x)\,\kappa))$$

$$\textit{where}$$
$$DoG(x, \kappa) \Longleftarrow \lambda v.G(x, DoR(v, \kappa)).$$
$$DoR(x, \kappa) \Longleftarrow \lambda v.\kappa h(x, v).$$

In words, G "sends" c to its continuation κ if $p(x)$ is true. Otherwise it computes $g_0(x)$ with a modified continuation, DoG. The new future of G is to save the result of the present computation while G computes $g_1(x)$. DoR applies h to the two results before resuming the original continuation.

These continuations inherit values from the present and record obligations for the future. They express the qualities of a control stack in a form suitable for reasoning. Wand (1980a) suggests that when seeking ways to implement recursion it is preferable to look for ways to represent continuations rather than to search a catalog of stack optimizations.

2.6.3. Distributivity of the Conditional, Revisited.
In Section 2.4.2 conditional selection was allowed to distribute over operands. By the following sleight-of-hand, we can conclude that distributivity applies to operations as well:

$$
\begin{aligned}
p \rightarrow f(x),\ g(y) &= p \rightarrow apply(f,\ x),\ apply(g,\ y) & \quad & \Delta\ apply \\
&= apply(\ (p \rightarrow f,\ g)\ ,\ (p \rightarrow x,\ y)\ ) & \quad & distributivity \\
&= (p \rightarrow f,\ g)\ (p \rightarrow x,\ y) & \quad & \Delta\ apply
\end{aligned}
$$

It is by no means clear that *apply* has a concrete counterpart in the typed specification language, for this would imply that operations exist that produce operations as values. The assumption of "functional operations", strains intuitive correlation between the underlying type and the designer's component catalog, much more so than the admission of structural combination. However, we shall see in Chapter 5 that this factorization can be meaningfully interpreted as a metaphor for communication.

2.6.4. Multiple Valued Functions.

To describe circuits, we must eventually deal with objects that have several outputs. Our development extends routinely to permit multiple valued operations, as we have already done by introducing parallel combination. However, the presence of multiple valued operations can lead to considerable confusion in detail. Whereas before we might appeal to the rigidity of the term translator τ (Prop. 2.4-1) to guarantee that all the arities match correctly, it now becomes necessary to keep track of arities explicitly. For example, if only one coordinate of a many valued operation is used, a projector must be introduced to access it. This matter of "typing" arises in several guises in the course of this presentation.

3. The Realization Language

A digital circuit description has two principal properties to specify: what components are in the circuit and how they are connected to each other. A number of assumptions are made concerning the nature of components. They are perceived as having physically distinct inputs and outputs. As a direct consequence, the model defined below cannot address such issues as the relational (as opposed to functional) qualities of fundamental electronic elements[1], and the bidirectional use of signal paths often found in physical implementations. The model is expressed in a language that describes logical behavior and physical connectivity, but not physical requirements such as power supply. Most important, the notation does not refer directly to timing. Component behavior is coordinated by storage elements called *registers* whose behavior in turn is governed by an external synchronizing agent or *clock*.

Since a component's inputs and outputs are distinct, its connectivity can be described by an applicative expression. The realization language is built from *signal expressions*, which are terminal terms that are sometimes annotated with an initialization clause. A signal expression denotes a *signal*, or history of values acquired over discrete time. That is, a signal is a non-terminating sequence of "instantaneous" values, and is modeled by the domain $Sig_D = D \times Sig_D$. The semantics of a circuit description will eventually be given as a fixed point in this

---

[1]A resistor is a constraint, such as $5\Omega = OHM \cap \{(v, i, r) \mid r = 5\}$ where $OHM = \{(v, i, r) \mid v = i * r\}$. To introduce a resistor as a component, one would have to choose between $5\Omega = (\lambda v.5 * i)$ and $5\Omega = (\lambda i.v \div 5)$. That is, either current or voltage would have to be free in a description involving resistors.

domain. However, the first concern is not so much with signals as with the values that occur on them. In particular we would like to know whether a signal produces a *specified* value at some time. Hence, it is appropriate at the outset to invoke the ordinary interpretation of a sequence as a function from time to values. In Sections 3.1 through 3.4 a signal expression is defined to denote such a function; in Section 3.5 the meaning of the realization language is restated in terms of sequences, where the obvious coercion, $behavior\!:\!Sig_D \rightarrow (Int \rightarrow D)$, relates the alternate semantics.

Under the functional interpretation, a realization defines a first order linear recurrence, a conventional formalism for digital behavior (See for example Hill and Peterson, 1968, Sec. 9.7). We shall later settle on the sequential interpretation because it leads to an experimental basis for design synthesis. Section 3.5 is a prelude to the implementation of realizations in the modeling language presented in Chapter 4.

3.1 Digital Circuit Descriptions

The computational aspects of a circuit are denoted by a set of components whose instantaneous behavior is that of an operator in some type.

DEFINITION 3.1-1. A combinatorial component *is an operator or predicate symbol in an underlying type.*

The symbol " ! " is reserved to denote storage in a manner described below. Storage components are informally called *registers*, although this term should not be taken literally. We shall build realizations from a language of signals, which express the behavior of components or groups of components.

DEFINITION 3.1-2. The language L_S of signal expressions contains terminal terms and terms of the form "c ! S", where c is a constant and S is a signal expression.

For the next two sections, components are enclosed in boxes to distinguish them from ordinary operators. Thus $\boxed{\text{add}}$ is a component and $\boxed{2}$ is a primitive signal expression over the integers. *Behavior,* defined just below, is a mapping from signal expressions and integers to values. That is, given a signal expression and a "time", behavior is the value on the expressed signal at that time. The symbol "@" is abbreviates this relation.

DEFINITION 3.1-3. Let L_S be the language of signal expressions over a type with carrier D; and let ω denote the non-negative integers. The function $*^{@*}:L_S \times \omega \to D$ defines the behavior *of ground term $s \in L_S$ at any time n as*

    i. $\boxed{c}^{@n} = c$, *for all n, where $c \in D$.*

    ii.   *If $c \in D$ and $s \in L_S$*

$$[c \, ! \, s]^{@0} = c, \text{ and}$$
$$[c \, ! \, s]^{@(n+1)} = s^{@n}.$$

    iii.   *If f is an m-place operation and $s_1, ..., s_m$ are in L_S, then for all n,*

$$[\boxed{f}(s_1, ..., s_m)]^{@n} = f(s_1^{@n}, ..., s_m^{@n})$$

Definition 3.1-3 accounts only for ground expressions. The behavior associated with an identifier is defined by equation and a circuit description is a system of equations.

DEFINITION 3.1-4. A signal equation *has the form "X = S" where X is an identifier and S is a signal. X* satisfies its defining equation *if and only if $X^{@n} = S^{@n}$ for all n. A* circuit description *is a system of signal equations, each defining a unique signal name.*

Identifiers are capitalized in circuit descriptions since they have become the serious symbols. The equality symbol denotes *behavioral equivalence,* or equality of value at all times, which is obviously an equivalence relation. Consider the circuit description

54

$$X = 1 \; ! \; \boxed{mpy}(Y, X)$$
$$Y = 1 \; ! \; \boxed{add}(\boxed{1}, Y)$$

It can easily be shown by induction that for all n, $Y^{@n} = n+1$ and $X^{@n} = n!$.

A circuit description is a linear form of circuit schematic. The identifiers name component outputs, and the equations specify connectivity. The description above is expressed graphically as

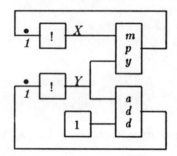

Schematics like the one above serve as an informal notation and will always be accompanied by a equational circuit description. The component $\boxed{!}$ is a generic clocked register, but reference to the clock is omitted. The tokens "$\bullet$" assert that at "time zero" the circuit is in a state where the registers contain the indicated values.

3.2 Translation to Circuit-Description Form

The central result of this chapter is that the correspondence between operations and components extends in a natural way to a correlation between the terms of an iterative specification and the signals of its realization. The change in interpretation of a term from instantaneous value to a behavioral counterpart is called *lifting*. It is of course no accident that circuit descriptions are in the same language of terms as was used to develop recursion schemes. Our first goal is to establish a relationship between the universal iterative scheme U_I and its register transfer counterpart. Let us establish some preliminary facts.

LEMMA 3.2-1. For constant a, signal S, and 1-place operation f,

$$\boxed{f}(a \ ! \ S) = f(a) \ ! \ \boxed{f}(S).$$

PROOF: By definition 3.1-3,

$$[\boxed{f}(a \ ! \ S)]^{@0} = f(a) = [\ f(a) \ ! \ \boxed{f}(S) \ ]^{@0}$$

and

$$[\boxed{f}(a \ ! \ S)]^{@(n+1)} = f([a \ ! \ S]^{@(n+1)}) = [\boxed{f}(S)]^{@n} = [f(a) \ ! \ \boxed{f}(S)]^{@(n+1)}$$

$\square$

LEMMA 3.2-2. Let f be a 1-place operation and let

$$X = a \ ! \ \boxed{f}(X)$$

$$U = a \ ! \ V$$

$$V = \boxed{f}(U)$$

Then X is behaviorally equivalent to U.

PROOF: (by induction on n)   Definition 3.1-3 shows $X^{@0} = a = U^{@0}$. Suppose that $X^{@n} = U^{@n}$. Then by Definition 3.1-3 and induction,

$$X^{@(n+1)} = f(X^{@n}) = f(U^{@n}) = V^{@n} = U^{@(n+1)}$$

$\square$

The connection between the specification and realization languages is made on the basis of the universal iterative scheme U_I.

56

THEOREM 3.2-3. Let F be defined by the recursion scheme

$$F(x) \Longleftarrow p(x) \to f(x), F(g(x)).$$

and let X be defined by the signal equation

$$X = \quad a \,! \; \boxed{g}(X)$$

Then if F converges on a, there is an n for which the following three statements hold:

   i. $p(X^{@n})$ *is true.*

   ii. $p(X^{@k})$ *is false for all $0 \le k < n$.*

   iii. $f(X^{@n}) = F(a)$.

PROOF: by subgoal induction on *F*. If *p(a)* is true then (*i–iii*) hold for *n = 0*. Otherwise, *F(a) = F(g(a))*. By induction hypothesis, there is an *N* such that (*i–iii*) hold for

$$V = g(a) \,! \; \boxed{g}(V)$$

By Lemma 3.2-1 $V = \boxed{g}(a \,! \; V)$. If *p(a)* is false, statements (*i–iii*) hold for the signal

$$U = a \,! \; V$$

when *n = N+1*. By Lemma 3.2-2, *X = U* and therefore statements (*i–iii*) also hold for *X*.

                                                              □

    Theorem 3.2-3 affirms the assertion made in Chapter 1 that the universal iterative scheme U_I is related directly to the universal register transfer schematic.

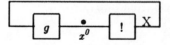

Here x^0 stands for the appropriate initial value on the signal *X*. We shall not address the question of how registers are initialized. To the basic feedback loop we may add two additional components, one to compute the terminal call, and one to represent the predicate. The component $\boxed{f}$ eventually produces a value equal to the specified function's result; the component $\boxed{p}$ produces a signal that indicates when a result is available.

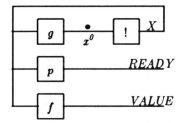

Expressed as a circuit description, this schematic translates to

$$X = x^0 ! \boxed{g}(X)$$

$$VALUE = \boxed{f}(X)$$

$$READY = \boxed{p}(X)$$

Since circuits do not "converge" to values, but rather "arrive at" them, we shall require a circuit that meets its specification to contain a signal that indicates when the specified value is present.

DEFINITION 3.2-4. A circuit description realizes a specification if and only if it has a signal READY, that states when the specified value is present, and VALUE, that contains the specified value.

3.3 Decomposition of Combined Components

Since any linear specification can be transformed to an instance of U_I, the circuit description above is a realization for all iterative specifications so long as combined operations are allowed in the underlying type. However, it does not have a very informative schematic; we should like to know what is going on inside those boxes. When structural combination was introduced in Section 2.4.3 it was claimed that it "made sense" in terms of circuit connectivity. We shall now justify that claim by showing that the packaging of operations in combined form is transparent to (*i.e.* distributes over) lifting. A given instance of the universal circuit description can always be decomposed into a more detailed schematic by reversing the transformation steps that combined operations.

To expose more details about the inner workings of a circuit, we must extract signals corresponding to individual registers within the state. If a combined operation is decomposed according to the propositions that follow, it is reduced to base terms that are either constants or projections. If projections are replaced by the identifiers from which they originated, they name signals tracing state-coordinate behavior. Consider a specification with formal parameter list *(u, v)*, that has been translated to an instance of U_p, and is therefore defined over a "monolithic" state identifier *Z*. Intuitively, if the behavior of *Z* is a sequence of pairs:

$$\left(\begin{pmatrix} u \, 0 \\ v \, 0 \end{pmatrix} \begin{pmatrix} u \, 1 \\ v \, 1 \end{pmatrix} \; . \; . \; . \;\right)$$

Z is decomposed by applying a lifted projector.

$$U = \boxed{\pi_1}(X)$$
$$V = \boxed{\pi_2}(X)$$

The following propositions establish the "local" transparency of structural combination to lifting. For serial combination, by Definition 3.1-3, for all *n*,

$$[\boxed{f \circ g}(Z)]^{@n} = (f \circ g)(Z^{@n}) = f(g(Z^{@n})) = f([\boxed{g}(Z)]^{@n}) = [\boxed{f}(\boxed{g}(Z))]^{@n}$$

Hence

PROPOSITION 3.3-1. $\boxed{f \circ g}(Z) = \boxed{f}(\boxed{g}(Z))$.

$\square$

Since signal-tuples have not been directly defined, there is no immediate correspondence between the forms like $\boxed{<f \, g>}(Z)$ and $(\boxed{f}(Z), \boxed{g}(Z))$. However, when serial and parallel combination are used in conjunction, as is always the case when generating combinations, the result is transparent.

PROPOSITION 3.3-2. $\boxed{f<g_1...g_m>}(Z) = \boxed{f}(\boxed{g_1}(Z),...,\boxed{g_m}(Z))$.

PROOF: By Proposition 3.3-1,

$$\boxed{f\langle g_1...g_m\rangle}(Z) = \boxed{f}(\boxed{\langle g_1...g_m\rangle}(Z)).$$

By Definition 3.1-3, the meaning of parallel combination, and for all n,

$$[\boxed{f\langle g_1...g_m\rangle}(Z)]^{@n} = \boxed{f}([\boxed{\langle g_1...g_m\rangle}(Z)]^{@n})$$

$$= f(\langle g_1 \cdots g_m\rangle(Z^{@n}))$$

$$= f(g_1(Z^{@n}),..., g_m(Z^{@n}))$$

$$= f([\boxed{g_1}(Z)]^{@n},..., [\boxed{g_m}(Z)]^{@n})$$

$$= [\boxed{f}(\boxed{g_1}(Z),...\boxed{g_m}(Z))]^{@n}$$

$$\square$$

By similar arguments,

PROPOSITION 3.3-3. $\boxed{\pi_i\langle g_1 \cdots g_m\rangle}(Z) = \boxed{g_i}(Z).$

$$\square$$

PROPOSITION 3.3-4. For any constant c, $\boxed{K^c}(Z) = \boxed{c}$.

$$\square$$

To deal with identifiers, we must look at lifting in the context of a defining equation. Consider the recursion scheme

$$F(x_1,..., x_n) \Leftarrow p \to r, F(t_1,..., t_n).$$

where propositional expression p, expression r, and all t_i are trivial. Under translator τ, F's transformation to an instance of U_1 yields

$$G(z) \Leftarrow \tau[\![p]\!](z) \to \tau[\![r]\!](z), G(\langle\tau[\![t_1]\!] \cdots \tau[\![t_n]\!]\rangle(z)).$$

The identifier z is now understood to name the state descriptor $(x_1,..., x_n)$. Our goal is to conclude that for any trivial term t, $\boxed{\tau[\![t]\!]}(Z) = t$. where the interpretation of t on the right is, of course, lifted. By Proposition 3.3-4, this equation holds if t is a constant; by Propositions 3.3-2 and 3.3-3, and structural induction over the language of terms, behavioral equivalence holds if t is a trivial application. The only remaining question is whether for identifier x, $\boxed{\tau[\![\, x_i \,]\!]}(Z) = X_i$. Since $\tau[\![\, x_i \,]\!] = \pi_i$, we must have $\boxed{\pi_i}(Z) = x_i$ to support the induction. Let us therefore introduce a signal equation $X_i = \boxed{\pi_i}(Z)$ for each identifier in F's defining equation. Now by Theorem 3.2-2 the circuit realization for G defines

$$Z = z^0 \, ! \, \boxed{<\tau[\![\, t_1 \,]\!]...\tau[\![\, t_n \,]\!]>}(Z)$$

If Z is replaced by its defining equation in each X_i's signal definition, we get

$$
\begin{aligned}
X_i &= \boxed{\pi_i}(z^0 \, ! \, \boxed{<\tau[\![\, t_1 \,]\!]...\tau[\![\, t_n \,]\!]>}(Z)) \\
&= \pi_i \, (z^0) \, ! \, \boxed{\pi_i}(\, \boxed{<\tau[\![\, t_1 \,]\!]...\tau[\![\, t_n \,]\!]>}(Z)) \qquad \text{\textit{Lemma 3.2-1}} \\
&= \pi_i \, (z^0) \, ! \, \boxed{\tau[\![\, t_i \,]\!]}(Z) \qquad\qquad\qquad \text{\textit{Props. 3.3-1 to 3.3-4}} \\
&= x_i^0 \, ! \, t_i \qquad\qquad\qquad\qquad\qquad\quad \text{\textit{by the argument above}}
\end{aligned}
$$

This establishes the following:

THEOREM 3.3-5. Let p be a trivial propositional expression. Let r, t_1 ,..., t_n be trivial expressions. The iterative recursion scheme

$$F(x_1 ,..., x_n) \Leftarrow p \to r, F(t_1 ,..., t_n).$$

is realized by the circuit equation

$$X_1 = x_1^0 \, ! \, t_1$$

$$\cdot$$
$$\cdot$$
$$\cdot$$

$$X_n = x_n^0 \, ! \, t_n$$
$$READY = \quad p$$
$$VALUE = \quad r$$

That is, if the registers that produce signals X_1 ,..., X_n are initialized with x_1^0 ,..., x_n^0 respectively, VALUE will contain $F(x_1^0 ,..., x_n^0)$ the first time READY is true. $\square$

3.4 Circuit Synthesis

We shall call instances of the scheme in Theorem 3.3-5 *simple loops*. A realization is obtained immediately from any simple loop by a transcription to circuit description form. The transcription indicates a change in the interpretation of the terms in the specification; they have been lifted. Because its realization is immediate, a principal method of circuit synthesis will be to find a simple loop version of a specification.

Let us bring our examples up to date. For concrete underlying types, the component counterpart of an operator or predicate will henceforth be written in upper case. Thus, *ADD* now denotes $\boxed{add}$.

Factorial. The initial specification for the *factorial* function was

$$FAC(x) \Leftarrow zero?(x) \to 1,\ mpy(x,\ FAC(dcr(x))\ ).$$

By Corollary 2.3-4 the simple loop

$$G(x,\ y) \Leftarrow zero?(x) \to y,\ G(dcr(x),\ mpy(x,\ y)).$$

gives the same answer when y is initialized to *1*. That is, $FAC(x^0) = G(x^0,\ 1)$ for all non-negative x^0. G's defining equation translates to the circuit description

$$X = x^0\,!\ DCR(X)$$
$$Y = \quad 1\,!\ MPY(X,\ Y)$$
$$READY = \qquad ZERO?(X)$$
$$VALUE = \qquad Y$$

By Theorem 3.3-5 the first time *true* appears on the *READY* signal, *VALUE* will contain $(x^0)!$.

Fibonacci. The initial specification was

$$FIB(x) \Leftarrow lt?(x,\ 2) \to 1,\ add(\ FIB(dcr(dcr(x))),\ FIB(dcr(x))\ ).$$

By Corollary 2.3-2 an equivalent simple loop is

$$G(x,\ y,\ z) \Leftarrow lt?(x,\ 2) \to y,\ G(\ dcr(x),\ z,\ add(y,\ z)\ ).$$

That is, for all x^0, $FIB(x^0) = G(x^0,\ 1,\ 1)$. Hence, a circuit that computes the *Fibonacci* function is described by

$$X = x^0 \text{ ! } DCR(X)$$
$$Y = 1 \text{ ! } Z$$
$$Z = 1 \text{ ! } ADD(Y, Z)$$
$$READY = ZERO?(X)$$
$$VALUE = Y$$

When X arrives at zero, Y will contain $FIB(x^0)$.

Greatest Common Divisor.   We began with

$$GCD(x, y) \Leftarrow eq?(x, y) \rightarrow x,$$
$$lt?(x, y) \rightarrow GCD(x, sub(y, x)), \ GCD(y, sub(x, y)).$$

The $lt?$-test must be distributed to get a simple loop. Since the specification is linear, the test can be implemented with a multiplexor:

$$G(x, y) \Leftarrow eq?(x, y) \rightarrow x, \ G\big(mux(lt?(x, y), x, y),$$
$$mux(lt?(x, y), sub(y, x), sub(x, y)) \big).$$

Arbitrarily push the conditional once more, inside the call to sub. We notice below a common subexpression that results.

$$G(x, y) \Leftarrow eq?(x, y) \rightarrow x, \ G\big(mux(lt?(x, y), x, y),$$
$$sub(mux(lt?(x, y), y, x),$$
$$mux(lt?(x, y), x, y)) \big).$$

This leads to the realization

$$X = x^0 \text{ ! } MUX(LT?(X, Y), X, Y)$$
$$Y = y^0 \text{ ! } SUB(MUX(LT?(X, Y), X, Y), MUX(LT?(X, Y), Y, X))$$
$$READY = EQ?(X, Y)$$
$$VALUE = X$$

$VALUE$ contains $GCD(x^0, y^0)$ as soon as $READY$ arrives at *true*. Of course, common subexpressions can be identified:

$$X = x^0\;!\;\; U$$
$$Y = y^0\;!\;\; SUB(U,\,W)$$
$$U = \qquad MUX(V,\,X,\,Y)$$
$$W = \qquad MUX(V,\,Y,\,X)$$
$$V = \qquad LT?(X,\,Y)$$
$$READY = \qquad EQ?(X,\,Y)$$
$$VALUE = \qquad X$$

A schematic for the *GCD* realization can be drafted from its signal equations:

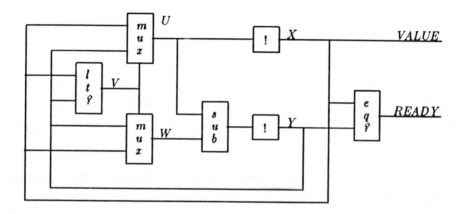

3.5 A Domain Model of Behavior

In this section the behavior model is restated in the Scott-Strachey notation. The motive for the translation will become apparent in Chapter 4 where an interpreter is presented for a version of the metalanguage. The restatement not only unifies the semantics of specifications and realizations, but will be used later as an interpretable basis for experimentation.

Signals have already been described informally as "value sequences", and it is clearly appropriate to model them as such. Let D be a flat domain of values that a signal can hold. A signal is in the domain of infinite sequences

$$Sig_D = D \times Sig_D$$

For $d \in D$, the constant signal $\boxed{d}$ is modeled as $d^\infty = fix \, \lambda s.\langle d \, , \, s \rangle$. Given $s \in Sig_D$ and $d \in D$, the register $d \, ! \, s$ is expressed simply as the pair $\langle d \, , \, s \rangle$.

Behavior is given by a function $behavior{:}Sig_D \to Int \to D$, where Int is the domain of integers (Sec. 2.5.1).

$$behavior = \lambda sn.(n = 0) \to (s{\downarrow}0), \, behavior(s{\downarrow}1)(n - 1)$$

By the fixed point property and for all n

$$behavior(d^\infty, n) = d,$$

$$behavior(\langle d, s \rangle, \, 0) = d \text{ and } behavior(\langle d \, , \, s \rangle, \, (n+1)) = behavior(s, n).$$

Hence, cases (i) and (ii) in Definition 3.1-3 are satisfied by the model. A 1-place component must give rise to a function in $Sig_D \to Sig_D$. However, we shall not take components themselves to be those functions, but instead define a component to be an operation-valued signal. That is, $Com_D = Sig_{D \to D}$; for instance, $\boxed{f} = f^\infty$. Until Chapter 5 only these constants in Com_D will be needed.

Application is generalized to deal with signals. Component application becomes coordinate-wise application of instantaneous operations to instantaneous values. For 1-place operations an $apply$-lifting functional like,

$$maplist = \lambda fs \, . \, \langle apply(f{\downarrow}0)(s{\downarrow}0) \, , \, maplist(f{\downarrow}1)(s{\downarrow}1) \rangle.$$

suffices. If the goal were to be a formal study in this model, it would be best to assume that all operations are unary, or perhaps $curry'd$. However, since our purpose remains to provide a notation that depicts implementations, we shall introduce a mechanism that admits n-ary (and n-valued) operations. There is a problem with structures: application of an n-placed component to n signals cannot be achieved by a simple mapping functional like $maplist$, for at each instant the operation expects its argument to be a tuple. An additional combinator is needed that, in effect, transposes a tuple-of-signals into a signal-of-tuples. If $f{:}D^n \to D$, the combinator needed to apply f^∞ is

$$transpose_n = \lambda s_1 ... s_n.\langle \langle (s_1{\downarrow}0) \, ,..., \, (s_n{\downarrow}0) \rangle \, , \, (transpose_n(s_1{\downarrow}1)...(s_n{\downarrow}1)) \rangle.$$

A generalized combinator, $transpose$, can be written to handle all dimensionalities. We arrive at the following definition of component application which we will denote with an infix colon. If $f \in Sig_{D^n \to D}$ and $s \in Sig_{D^n}$ then

$$[\![\, f : s \,]\!] = maplist\ f\ (transpose\ s).$$

In the domains of signals and components, the meaning of a circuit description can now expressed as the fixed point. That is, circuit description $\{X_i = S_i\}$ has meaning

$$fix\ (\lambda\ [\ X_1, ..., X_n\ ] \cdot \langle S_1, ..., S_n \rangle\ )$$

This model of behavior is essentially that of Kahn (1973) who also uses an equational signal definition style. It is easily related to Milner's simple process behavior model (1973) as presented by Gordon (1980). Minor variations are due to differences in emphasis. Milner's model is more generally descriptive. He defines the domain of *processes* as follows:

$$Process = Input \to (Output \times Process)$$

That we have modeled components instead as higher order signals is a technical point, since only constants in Com_D are permitted. This restriction is relaxed slightly and only temporarily in Section 5.1, as a means to introduce communication. Milne and Milner (1979) present an algebra of connectivity that covers a wider class of concurrent behavior than is attempted here. This point is discussed further in Chapter 7.

4. DAISY

DAISY is an interpreted language in which both specifications and realizations can be implemented. It is a descendant of Pure LISP (McCarthy, *et.al*, 1965) and to a lesser extent of SCHEME (Sussman and Steele, 1978). Its interpreter executes in a data space of binary list cells and uses graph reduction to solve recursive equations. DAISY's syntax is similar to many contemporary *applicative* languages (Burge, 1975) (Henderson, 1980); it is a language of expressions with no explicit sequential control constructs. Computation is *demand driven*, making interpretation yield "call-by-name" semantics. Consequently, specifications in DAISY are entirely consistent with the valuation function of Definition 2.2-2. Moreover, circuit descriptions can also be computed even though representation of behavior involves infinite data structures. We will take a brief look at DAISY's implementation and then give a formal definition of a subset of the language. The remainder of this chapter is devoted to demonstrating how DAISY might be used to support circuit synthesis.

4.1. Operational Semantics – a Summary

Functional language interpreters can be classified in terms of string reduction, although few actually work that way. Instead, they use graph reduction to reduce duplication of substituted text. The necessary bookkeeping is implemented by a hidden data structure called an *environment*, which represents a mapping from identifiers to values (see Sec. 2.2). Substitution steps are emulated by adding new bindings for formal parameters in this data structure.

Recall from Definition 2.2-2 that the value of an expression depends in part on the substitution of actual arguments for formal parameters according to function definitions. In reasoning about reduction we could arbitrarily unfold serious terms. However, mechanical evaluators must have a *computation rule* by which they deterministically select which terms to unfold. A leftmost-innermost rule is most often used: the interpretation algorithm unfolds the first iterative term it encounters reading left to right, makes a substitution, then simplifies. This reduction strategy is referred to as *call-by-value* interpretation, since it mimics that operational argument evaluation protocol as defined in ALGOL 60 (Backus, *et. al.*, 1963).

DAISY's computation rule is leftmost-outermost, meaning that the corresponding string reduction interpreter expands the first function variable symbol it encounters. This is analogous to passing text rather than values to subprograms and so is called a *call-by-name* computation rule.

The advantage of call-by-value is its relative efficiency on conventional architectures. However, call-by-name is a stronger rule: it produces results more often. The difference is illustrated by a simple example. Consider the system

$$F(x,\ y) \Longleftarrow x.$$

$$G(x) \Longleftarrow F(x,\ G(x)).$$

and suppose that the ground term $G(a)$ is to be evaluated. Let $\xrightarrow{F}$ and $\xrightarrow{G}$ indicate reduction according to the definitions of F and G respectively. The reduction sequences under the two computation rules are

$$G(a) \xrightarrow{G} F(a,\ G(a)) \xrightarrow{F} a \qquad\qquad \textit{(call-by-name)}$$

$$G(a) \xrightarrow{G} F(a,\ G(a)) \xrightarrow{G} F(a,\ F(a,\ G(a))) \xrightarrow{G} ... \qquad \textit{(call-by-value)}$$

This reduction clearly diverges under the call-by-value computation rule. If underlying operations are assumed to be strict, then call-by-name interpretation converges whenever a value is defined (Manna 1974, p. 388).

DAISY inherits its computation rule from the mechanisms it uses to manipulate its data space. When a new record is built, each of its fields is filled with a *suspension,* or expression closure, which contains the information needed to compute the value of that field. The computation does not take place unless and until the field is accessed. Once access occurs, and if the suspension converges, the referent field is updated with the result, so that subsequent access need not recompute it. This basic model of computation has many names including *lazy evaluation* (Henderson and Morris, 1976), *delay rule* (Vuillemin, 1974), and *demand driven* (Ashcroft and Wadge, 1977; Kahn and MacQueen, 1977). The

last of these will be used here. Two consequences of demand driven computation
are of importance here:

- Since environments are suspended, argument evaluation is deferred until
  identifier bindings are sought. In the absence of other side effects, the
  deferment yields the call-by-name characterization of interpretation, with
  some improvement in efficiency because redundant reductions are shared
  (Friedman and Wise, 1976a).

- Non-finite data structures can be built from finite descriptions[1]. Only those
  portions of such structures that are needed are actually brought into being.
  (Friedman and Wise, 1976b, 1976c; Friedman, Wise, and Wand, 1976). In
  particular, the signals that are modeled as infinite sequences in Section 3.4
  can be readily expressed and manipulated in DAISY.

DAISY is a vehicle to state specifications and realizations in executable form.
Specifications are not compromised by the interpreter's evaluation strategy
because the call-by-name semantics are consistent with their formal meanings.
The facility to manipulate infinite objects allows logical descriptions of circuits to
be explored through direct emulation. If it is granted that the realization
language is an adequate starting point to fabricate an implementation, its direct
interpretation is a way to observe product behavior without building a physical
prototype.

4.2. The Language

Figure 4.1 gives an idealized definition of DAISY's syntax. (The parser for
this grammar has not been fully implemented; Appendix A gives a description of
the current syntax.) The stylized syntax is used from this point on in examples,
since it more closely reflects the notation we have developed so far. Actual
source for the running examples is shown in Appendix B. In the figure the alternate forms of *conditional* and *body* have the same meanings; which to use is a
matter of preference or style. For example, if *p*, *c*, and *a* are lexically small, it is
probably better to write "*p* → *c, a*" rather than " **if** *p* **then** *c* **else** *a*" since the
keywords in the second version visually dominate the text.

---

[1]This is a simplification since descriptions are themselves data. It is only required
that the description be finitely describable, and so on.

$$
\begin{aligned}
\textit{expression} \;&::=\; (\; \textit{expression} \;) \;\mid\; \text{@} \;\textit{expression} \;\mid\; \textit{atom} \\
&::=\; \textit{fern} \;\mid\; \textit{application} \;\mid\; \textit{abstraction} \;\mid\; \textit{conditional} \;\mid\; \textit{system}
\end{aligned}
$$

$$
\textit{atom} \;::=\; \textit{identifier} \;\mid\; \textit{numeral} \;\mid\; \textit{operator}
$$

$$
\begin{aligned}
\textit{fern} \;&::=\; [\,\textit{list}\,] \;\mid\; <\,\textit{list}\,> \;\mid\; \{\,\textit{list}\,\} \\
\textit{list} \;&::=\; \Lambda \;\mid\; \textit{expression} \; \bullet \;\mid\; \textit{expression}\,!\,\textit{expression} \;\mid\; \textit{expression list}
\end{aligned}
$$

$$
\textit{application} \;::=\; \textit{expression} : \textit{expression}
$$

$$
\textit{abstraction} \;::=\; \lambda \,\textit{expression} \,.\, \textit{expression}
$$

$$
\begin{aligned}
\textit{conditional} \;&::=\; \textbf{if}\;\textit{expression}\;\textbf{then}\;\textit{expression}\;\textbf{else}\;\textit{expression} \\
&::=\; \textit{expression} \rightarrow \textit{expression} \,,\, \textit{expression}
\end{aligned}
$$

$$
\begin{aligned}
\textit{system} \;&::=\; \textit{body} \;\mid\; \textbf{rec}\;\textit{body} \\
\textit{body} \;&::=\; \textbf{let}\;\textit{specification}\;\textbf{in}\;\textit{expression} \\
&::=\; \textit{expression}\;\textbf{where}\;\textit{specification} \\
\textit{specification} \;&::=\; \Lambda \;\mid\; \textit{definition specification} \\
\textit{definition} \;&::=\; \textit{expression} = \textit{expression} \\
&::=\; \textit{identifier} : \textit{expression} \Longleftarrow \textit{expression} \,.
\end{aligned}
$$

Figure 4.1. DAISY Expression Syntax.

Function definitions are similar to the notation of Chapter 2. The three example functions might be defined as follows in DAISY:

rec *Expression* **where**

FAC:x $\Longleftarrow$ zero?:x $\rightarrow$ 1, FAC:dcr:x.

FIB:x $\Longleftarrow$ **if** lt?:$<$x 2$>$ **then** 1
else add:$<$ FIB:dcr:dcr:x FIB:dcr:x $>$.

GCD:x $\Longleftarrow$ **let** [u v] = x
       in
          if eq?:x **then** u
          else if lt?:x **then** GCD:$<$u sub:$<$v u$>$ $>$
              else   GCD:$<$v sub:$<$u v$>$ $>$.

Expression would contain some ground expression to be evaluated according to this specification. We shall usually display specifications in the context of some "experiment" like this[2]. The somewhat contrived version of *GCD* illustrates DAISY's lack of emphasis on argument structure. Although *GCD* takes two arguments, its formal parameter does not name them. It is the inner specification, "**let** [u v] = x ..." that identifies x's coordinates. The "2-place" operations *eq?* and *lt?* can be applied directly to *x*, since it will have the required structure.

4.3. Formal Semantics of a Subset of DAISY

The language definition in this section omits some features of DAISY that are not used in this investigation. There is a construct for indeterminacy (*ferns* of the form { ... }) which has only recently been formalized (Wise, 1983). Operational discussions of this construct have been published by Friedman and Wise (1979, 1980, 1981) and Filman and Friedman (1983). As in LISP, expression text is indistinguishable from ordinary data in DAISY's data space, and programs can be written to produce other programs. However, DAISY's program representation is rather involved; discussion of it is omitted since program builders are not presented here.

The figures referred to in this discussion appear at the end of the section. Figure 4.2 is a simplified language that will be used for DAISY's formal definition. With the exceptions already mentioned, expressions in the full language can easily by converted to this "kernel" language. Some examples of the conversion are shown in Figure 4.3. Figure 4.4 gives a standard semantics for the kernel language.

Domains (Figure 4.4a). *Opr* is a set of identifiers reserved to denote primitive operations on DAISY's underlying type, *Val*. Some of DAISY's operators are summarized in Figure 4.5. The structure of formal arguments, given by the domain equation for *Arg,* comes into play in defining environment extension. Included in *Val* are the primitive syntactic types and a set of *messages* that are returned when expressions are found by the interpreter to be erroneous or meaningless. Operations also produce messages; for example, an arithmetic operation returns an error message on non-numeric operands. The non-flat summands of

---

[2]In the implementation, functions may be directly defined at top level as though the operator's programming environment had been initiated in a **rec-where**.

Val are *Cls,* a domain of function closures, and *Lst,* the domain of value pairs. *Env* is the usual domain of environments, that map identifiers to their bindings. The primitive valuations for numerals and operators are left unspecified.

Semantics   (Figure 4.4b). The interpreter is specified by the valuation function *ID,* with auxiliary combinators as defined in Figure 4.4c. Numerals, operators, and quoted identifiers evaluate to themselves; the empty *fern* evaluates to *Nil.* Unquoted identifiers evaluate to their bindings in the current environment. Value pairs are expressed by list concatenation. Abstractions are closed in the environment in which they are evaluated (making DAISY a lexically scoped language). Conditional expressions and recursive definitions have standard meanings. The interpretation of application is discussed below when the auxiliary combinator *d-apply* is introduced.

Auxiliaries   (Figure 4.4c). The environment extension combinator binds structures to values. The formal argument is used as a pattern by which the value is accessed; identifiers are bound to their corresponding locations. If the formal argument is a simple list, the effect is the same as a call-by-name parameter passing protocol. As the *GCD* example above indicates, the formal argument may be used to name arbitrary pieces of the actual argument. The implied principle is that all functions are monadic, and that formal arguments serve as a kind of record declaration. However, the interpreter does not check for a pattern match at binding time, as to do so would introduce strictness. A list membership operation, called *Member?,* might be defined[3]:

$$\text{Member?} : x \Leftarrow$$
$$\textbf{let } [\, a \, ! \, L \,] = x$$
$$\textbf{let } [\, e \, ! \, L' \,] = L$$
$$\textbf{in}$$
$$\quad \textbf{if } \text{null?} : L \textbf{ then } <>,$$
$$\quad \textbf{else if } \text{same?} : x \textbf{ then } @\text{true},$$
$$\quad \textbf{else } \text{Member?} : <\, a \, ! \, L' \,>.$$

The **let**-definition gives names to the components of the formal argument, *x.*

---

[3]This example, like the *GCD* definition, is meant to illustrate a point about binding in DAISY, and is not put forward as an example of good programming style!

The "binary" operation *same?* is applied to x because it happens that its first two elements are the ones that need to be compared. As is the case with all such operators in DAISY, *same?* does not require that its argument be of length two. The head and tail of the list L are named a and L', even though L might be empty. Again, this is valid in DAISY because there is an intervening *null?*-test before L' is used.

Application is orthogonal, meaning that the evaluator renders an interpretation for any value that appears in the function position. This is shown in the definition of auxiliary function *d-apply*. Numerals, for example, are taken to denote list *probes* returning the element at the appropriate coordinate of the argument. If the function-part is a list, its elements are applied coordinate-wise on the transposed argument. This choice of interpretation for list application comes out of the investigations by Friedman and Wise (1976c, 1978a) of systems programming, but it is also consistent with the circuit behavior model of Section 3.4.

The function *predicate* assigns an interpretation of truth to every value; as in LISP, *Nil* is the only instance of falsity. The Boolean interpretation of a message is erroneous. On valid values *predicate's* result is a branch-like operation that selects an alternative, by coercing one of the values in a pair. The reader can check that the conditional is non-strict in its alternatives.

We shall say no more about the implementation of DAISY except to note the important fact since the list construction primitive is suspending list concatenation is not a strict operation. Write "$x \equiv y$" to mean that $x = y$ and $x = \bot$ iff $y = \bot$. By the definition of $I\!D$, it is straightforward to show

PROPOSITION 4.2-1. For all environments ρ, and all expressions e and e',

$$I\!D [\![\ (\ \lambda \ [\ \mathbf{h} \ ! \ \mathbf{t} \ ] \ . \ \mathbf{h}) : \ < e \ ! \ e' > \ ]\!] \rho \equiv I\!D [\![\ e \ ]\!] \rho$$

and

$$I\!D [\![\ (\ \lambda \ [\ \mathbf{h} \ ! \ \mathbf{t} \ ] \ . \ \mathbf{t}) : \ < e \ ! \ e' > \ ]\!] \rho \equiv I\!D [\![\ e' \ ]\!] \rho$$

PROOF: (Appendix C).

$\square$

These strong equalities are maintained by DAISY's implementation. Hence, the pairing and projection functions of the Scott-Strachey language can be implemented by

$$pair : [\ x\ y\ ] \Longleftarrow \ <\ x\ !\ y >.$$

$$head : [\ h\ !\ t\ ] \Longleftarrow h\ .$$

$$tail : [\ h\ !\ t\ ] \Longleftarrow t\ .$$

The required axioms:

$$head\ :\ pair\ :\ <\ e\ \ e'\ >\ \equiv e$$
$$tail\ :\ pair\ :\ <\ e\ \ e'\ >\ \equiv e'$$

are satisfied in the implementation.

expression ::= **@** *identifier* | (*expression*)

          atom | *fern* | *application* | *abstraction* | *conditional* | *system*

atom ::= *identifier* | *numeral* | *operator*

fern ::= <> | < *expression* ! *expression* >

application ::= *expression* : *expression*

abstraction ::= λ *argument* . *expression*

argument ::= [ ] | *identifier* | [*argument* ! *argument*]

conditional ::= **if** *expression* **then** *expression* **else** *expression*

system ::= **rec** *argument* = *expression* **in** *expression*

Figure 4.2. DAISY's Kernel Syntax.

Let e and v be any expressions, i any identifier, and x any argument.

$$@e \quad - \quad \textit{not permitted unless e is an identifier}$$

$$\{...\} \quad - \quad \textit{not permitted}$$

$$\lambda\, x\,.\, e \quad - \quad \textit{not permitted unless x is an argument}$$

$$< e\ e'> \quad \longleftrightarrow \quad < e\,!<e'\,!<>>>$$

$$[\, x\ x'\,] \quad \longleftrightarrow \quad [\,x\,!\,[\,x'\,!\,[\,]\,]\,]$$

$$<e *> \quad \longleftrightarrow \quad (\lambda\, \mathbf{i}\,.\,\mathbf{rec}\,\mathbf{j} = < \mathbf{i}\ !\ \mathbf{j}> \mathbf{in}\ \mathbf{j}\,)\,:\,e$$

$$[\,i\,!\,i'\,] \quad \longleftrightarrow \quad <@i\ !\ @i'> \quad \textit{(N.B. As a value, only)}$$

$$F : x \Longleftarrow e\,. \quad \longleftrightarrow \quad F = (\,\lambda\, x\,.\, e\,)$$

$$e\ \mathbf{where}\ x = e' \quad \longleftrightarrow \quad (\lambda\, x\,.\, e)\,:\,e'$$

$$\mathbf{let}\ x = v\ \ x' = v'\ \mathbf{in}\ e \quad \longleftrightarrow \quad (\lambda\,[\,x\ x'\,]\,.\, e)\,:<v\,v'>$$

$$e \to e',\, e' \quad \longleftrightarrow \quad \mathbf{if}\ e\ \mathbf{then}\ e'\ \mathbf{else}\ e'$$

$$\mathbf{rec\ let}\ e \quad \longleftrightarrow \quad \mathbf{rec}\ e$$

Figure 4.3. Conversions to the Kernel Language.

<u>Syntactic Domains</u>

Ide	*(i)*	*identifiers*
Nml	*(n)*	*numerals*
Opr	*(o)*	*operator symbols*
$Arg = Nil + Ide + (Arg \times Arg)$	*(x)*	*formal arguments*
Exp	*(e)*	*expressions*

<u>Semantic Domains</u>

Int		*integers*
Nil		*nullary value*
$Opn = Val \rightarrow Val$		*operations*
$Val = Nil + Ide + Nml + Opr + Msg + Cls + Lst$	*(v)*	*values*
$Msg = \{\,\text{``Invalid function''}, \ldots\}$	*(m)*	*messages*
$Cls = (Val \rightarrow Val)$	*(f)*	*function closures*
$Lst = (Val \times Val)$	*(l)*	*lists*
$Env = Ide \rightarrow Val$	*(ρ)*	*environments*

<u>Valuations</u> (see Figure 4.4b)

$\mathbb{N}\!: Nml \rightarrow Int$	*Numeral meanings – unspecified*
$\mathbb{K}\!: Opr \rightarrow Opn$	*Operator meanings – unspecified*
$\mathbb{D}\!: Exp \rightarrow Val$	*Expression evaluation*

Figure 4.4a. DAISY's Standard Semantics – Domains.

$$\mathbb{D} : Exp \rightarrow Env \rightarrow Val$$

$$\mathbb{D}[\![\ n\ ]\!]\rho = n$$

$$\mathbb{D}[\![\ @\ i\ ]\!]\rho = i$$

$$\mathbb{D}[\![\ o\ ]\!]\rho = o$$

$$\mathbb{D}[\![\ <>\ ]\!]\rho = Nil$$

$$\mathbb{D}[\![\ i\ ]\!]\rho = \rho(i)$$

$$\mathbb{D}[\![\ <e_1!e_2>\ ]\!]\rho\ \langle\ \mathbb{D}[\![\ e_1\ ]\!]\rho\ ,\ \mathbb{D}[\![\ e_2\ ]\!]\rho\ \rangle$$

$$\mathbb{D}[\![\ e_1:e_2\ ]\!]\rho = \text{d-apply}\ (\mathbb{D}[\![\ e_1\ ]\!]\rho)\ (\mathbb{D}[\![\ e_2\ ]\!]\rho)$$

$$\mathbb{D}[\![\ \lambda\ x\ .\ e\ ]\!]\rho = \lambda v.\mathbb{D}[\![\ e\ ]\!]\ \rho[\ v\ /\ x\ ]$$

$$\mathbb{D}[\![\ \text{if}\ e_1\ \text{then}\ e_2\ \text{else}\ e_3\ ]\!]\rho = (predicate(\mathbb{D}[\![\ e_1\ ]\!]\rho))\ \langle\mathbb{D}[\![\ e_2\ ]\!]\rho\ ,\ \mathbb{D}[\![\ e_3\ ]\!]\rho\rangle$$

$$\mathbb{D}[\![\ \text{rec}\ x = e_1\ \text{in}\ e_2\ ]\!]\rho = \mathbb{D}[\![\ e_2\ ]\!]\ (\text{fix}\ \lambda\rho'\ .\ \rho[\ \mathbb{D}[\![\ e_1\ ]\!]\rho'\ /\ x\ ])$$

Figure 4.4b. DAISY's Standard Semantics – Valuation.

78

$$\textit{Environment Extension, (*[*/ *]): Env} \rightarrow \textit{Val} \rightarrow \textit{Arg} \rightarrow \textit{Env}$$

$$\rho[\ v\ /\ x\ ] = \lambda i.(x\ isNil) \rightarrow \rho(i),$$
$$(x\ isIde) \rightarrow [(x = i) \rightarrow v,\ \rho(i)],\ \rho[\ tl\ v\ /\ x\downarrow 1\ ][\ hd\ v\ /\ x\downarrow 0\ ]$$

$$\textit{d-apply: Val} \rightarrow \textit{Val} \rightarrow \textit{Val}$$
$$d\text{-}apply = \lambda fa.(f\ isMsg) \rightarrow \text{``Invalid Function''},$$
$$(f\ isIde) \rightarrow \text{``Undefined Function Symbol} - f\text{''},$$
$$(f\ isOpr) \rightarrow (\mathbb{K}\ f)(a),$$
$$(f\ isNil) \rightarrow Nil,$$
$$(f\ isNml) \rightarrow probe(\mathbb{N}\ f)a,$$
$$(f\ isCls) \rightarrow fa,$$
$$(f\ isLst) \rightarrow \langle\ d\text{-}apply(hd\ f)(hds\ a)\ ,\ d\text{-}apply(tl\ f)(tls\ a)\ \rangle.$$

$$\textit{predicate: Val} \rightarrow \textit{(Lst} \rightarrow \textit{Val)}$$
$$predicate = \lambda v.(v\ isMsg) \rightarrow (\ \lambda l.\text{``Bad proposition''}),$$
$$(v\ isNil) \rightarrow (\ \lambda l.l\downarrow 1)\ ,\ (\ \lambda l.l\downarrow 0)\ .$$

$$\textit{probe: Int} \rightarrow \textit{Val} \rightarrow \textit{Val}$$
$$probe = \lambda nl.(n = 0) \rightarrow (hd\ l),\ probe(n-1)(tll).$$

$$\textit{hd, tl, hds, tls: Val} \rightarrow \textit{Val}$$
$$hd = \lambda v.(v\ isLst) \rightarrow v\downarrow 0,\ \text{``Invalid hd-access''}.$$
$$tl = \lambda v.(v\ isLst) \rightarrow v\downarrow 1,\ \text{``Invalid tl-access''}.$$
$$hds = v.(v\ isNil) \rightarrow Nil,\ \langle\ hd(hdv)\ ,\ hds(tlv)\ \rangle$$
$$tls = v.(v\ isNil) \rightarrow Nil,\ \langle\ tl(hdv)\ ,\ tls(tlv)\ \rangle$$

Figure 4.4c. DAISY's Standard Semantics – Auxiliaries.

<u>Reference comparison</u>
same? – *reference equality*

<u>Type Predicates</u>

null? – *test for nil*	**not** – *not null*	**list?** – *non-null list*
nmbr? – *numeral?*	**ltrl?** – *literal atom?*	**atom?** – *numeral or literal*

<u>Numeric Comparisons</u>

zero? – *test for zero*

lt? – *less than*	**eq?** – *equal*	**gt?** – *greater than*
le? – *at most*	**ne?** – *not equal*	**ge?** – *at least*

<u>Unary Numeric Operators</u> *(Numbers are represented in Rational form)*

sgn – *sign (–1, 0, 1)*	**inc** – *increment*	**dcr** – *decrement*
neg – *negate*	**num** – *numerator*	**den** – *denominator*
inv – *invert*	**quo** – *quotient*	**mod** – *remainder*

<u>Binary Numeric Operators</u>

add – *addition*	**sub** – *subtraction*
mpy – *multipliction*	**div** – *division*

<u>Constructors and Probes</u>

cons – $(\lambda\ [h\ t]\ .\ <h\ !\ t>)$	**frons** – $(\lambda\ [h\ t]\ .\ \{h\ !\ t\})$
first – $(\lambda\ [h\ !\ t]\ .\ h)$	**rest** – $(\lambda\ [h\ !\ t]\ .\ t)$

<u>List Operators</u>

if – *extended conditional, as in* $(p_1 \rightarrow v_1,\ p_2 \rightarrow v_2 ,...,\ p_n \rightarrow v_n)$

in? – *list membership*

sigma – *numeral summation*	**pi** – *numeral product*
and – *all true*	**or** – *not all null*

<u>Input/Output</u>

console – *prompt-character $\rightarrow$ character-stream-from-keyboard*

screen – *character-stream $\rightarrow$ terminal display*

dski – *host-file $\rightarrow$ character stream*

dsko – *character-stream $\rightarrow$ host-file*

parse – *character-stream $\rightarrow$ expression-stream*

evlst – *expression-stream $\rightarrow$ value-stream*

issue – *value-stream $\rightarrow$ character-stream*

Figure 4.5. Some DAISY Operations.

4.4. Circuit Emulation

It has already been noted that specifications can readily be transcribed into DAISY to get executable versions. Realizations are just as easily transcribed, as is demonstrated later in this section. The formal model of components and signals as infinite sequences may be implemented by constructing infinite lists to represent them. Since dependencies among signals are well behaved, there is no difficulty in building and manipulating their representations. It is ultimately the need to observe these objects that brings them into existence. The manner in which the observation is made determines how computation takes place. After a discussion of this issue, we return to the earlier examples and observe their behavior through emulation in DAISY.

4.4.1. Non-finite Data Structures.   The function K, defined as

$$K : c \Longleftarrow \; <c \; ! \; (K : c) \; >.$$

produces a list whose head is c's value and whose tail is also such a list. In fact, for any positive numeral n, the expression $n : K : c$ returns c's value; the list is infinite for all intents and purposes. The definition

$$K : c \Longleftarrow \mathbf{rec} \; L \; \mathbf{where} \; L = \; <c \; ! \; L>.$$

yields the same result since it specifies that L must be a list whose head is the value of c and whose tail is also such a list. There is a special symbol in DAISY to express constant sequences like this. One may define

$$K : c \Longleftarrow \; <c \; *>.$$

The asterisk is meant to be suggestive of a Kleene star, and should be taken to mean "arbitrarily many c's." The function N, defined as

$$N : c \Longleftarrow \; <c \; ! \; (N : add : \; <c \; 1>) \; >.$$

produces a list of increasing numerals. N's value can also be described by a "data recursion"

$$N : c \Longleftarrow \mathbf{rec} \; L$$
$$\mathbf{where}$$
$$L = \; <c \; ! \; (<add \; *> : < \; <1*> \; L \; >) \; >$$

Let $c = 2$. The computation of $<add \; *> : < \; <1 \; *> \; L \; >$ can be pictured as a progressing sum, with each element of L resulting from the previous value:

$$< 1 * > = \quad 1 \ 1 \ 1 \ 1 \ 1 \ 1 \ ...$$
$$L = \quad 2 \ 3 \ 4 \ \underline{\phi} \ \phi \ \phi \ \phi \ ...$$

$$< \text{add} * > = \quad 3 \ 4 \ \underline{5} \ \phi \ \phi \ \phi \ \phi \ ...$$

The translation from circuit description to DAISY expression is straightforward. It follows the model of component behavior defined in Section 3.4:

signal or component	behavior semantics	DAISY expression
$\boxed{c}$	c^{∞}	$< c * >$
$\boxed{f}(t_1 ,..., t_n)$	$(f^{\infty}) : \langle t_1 ,..., t_n \rangle$	$< f * > : < t_1 \ ... \ t_n >$
$c \ ! \ t$	$\langle c, t \rangle$	$< c \ ! \ t >$

We will retain the convention of using operator symbols written in upper case to refer to components. The example above would define *ADD*, to be "$<$add $*>$". However, an adjustment is needed to make components out of DAISY's monadic operations, which are applied directly to their operands and not to 1-tuples. For example, one writes[4] "inc : n" rather than "inc : $<$n$>$" to increment a numeral. This does not fit with the usual transposition. Although one is tempted to write "$<$inc $*> : <5 *>$" to get a stream of 6's, the argument $<5 *>$ cannot be transposed because it is not a stream of tuples. On the other hand, the expression $<<5 *>>$ transposes to $<<5> *>$ to get a uniform stream of 1-tuples. To increment this stream, *inc* should expect an argument-list of length one. Thus, the component version of a monadic operation like *inc* is defined as

$$\text{INC} = < (\lambda \ [\ n \ ] \ . \ inc : n \ ) \ * >$$

Figure 4.6 gives component versions of the DAISY operations used later.

---

[4]so as to avoid expressions like "inc $: <$inc $: <$inc$:5>>>$".

$$\begin{aligned}
\text{ADD} &= <\text{add} *> \\
\text{DCR} &= < (\lambda\,[\,x\,]\,.\,\text{dcr} : x\,) *> \\
\text{DIV} &= <\text{div} *> \\
\text{INC} &= < (\lambda\,[\,x\,]\,.\,\text{inc} : x\,) *> \\
\text{LT?} &= <\text{lt?} *> \\
\text{MPY} &= <\text{mpy} *> \\
\text{MUX} &= <\text{if} *> \\
\text{SUB} &= <\text{sub} *> \\
\text{ZERO?} &= < (\lambda\,[\,x\,]\,.\,\text{zero?} : x\,) *>
\end{aligned}$$

Figure 4.6. DAISY Component Implementations.

4.4.2. Output Driven Computation. In a purely demand driven model, computation is caused by the need for a result. Ultimately, need is determined by the device that displays that result. One can build and manipulate non-finite data structures in DAISY as long as care is taken about how they are displayed (Friedman and Wise, 1976b). This relationship can be specified by introducing a causal operation called *strict* whose convergence depends on the existence of a value. Thus, *strict* : $<u\ v>$ returns v's value after u has converged[5].

Through judicious use of *strict*, call-by-value interpretation can be imposed in DAISY. For example, function F defined by F : [n m] $\Longleftarrow$ e becomes call-by-value[6] when transformed to

$$F : [\,n\ m\,] \Longleftarrow strict : <\ n\quad strict : <\ m\ e\ >>.$$

To address the relationship between input/output and computation let us define a *device* to be a manipulator of atom streams. A single occurrence of *strict* is used to define the behavior of a atom stream consumer:

---

[5]Note that u is not necessarily made fully manifest. For example, *strict* : $<<u\,!\,u'\,>\,v>$ converges independent of u and u' because the list constructor is not strict.

[6]Conversely, in call-by-value interpreters function closures can be used to induce call-by-name through a dual operation called *delay* (Landin, 1965; Henderson, 1980).

$$\text{Display} : [\, c\, !\, S\,] \Longleftarrow \textit{strict} : <\, c\, !\, (\text{Display} : S)\, >.$$

We shall make no assumption of temporal order in defining a generic input device:

$$\text{Receive} : x \Longleftarrow <\, \textit{random-atom}\, !\, (\text{Receive} : x)\, >.$$

Ideally it is Display that brings characters forth in the order they are typed. Imagine a program that converts values into atom streams:

$$\text{Print} : L \Longleftarrow \textbf{rec } \text{Help} : <L\, <>>.$$

> **where**
>
> $\text{Help} : [\, L\, X\,] \Longleftarrow$
>   **let** $[\, u\, !\, v\,] = L$
>   **in**
>   **if** atom? : L **then** $<L\, !\, X>$ **else** Help : $<u$ Help : $<v\, c>\, >.$

Now consider the expression Display : Print : $<e\, !\, e'>$. The computation is ordered, first e then e', because Print produces its stream that way and Display consumes the stream in order. If e' diverges, the prefix of the result is still displayed. In fact, the computation of e' does not take place until after e's value has been transmitted. We shall make use of this fact when we attempt to observe circuits in emulation.

4.4.3. Experimentation with Realizations.

Recall that the *factorial* specification is realized by the circuit description

$$
\begin{aligned}
X &= x^0\ !\ \text{DCR}(X)\\
Y &=\ 1\ !\ \text{MPY}(X, Y)\\
\text{READY} &=\quad \text{ZERO?}(X)\\
\text{VALUE} &=\quad Y
\end{aligned}
$$

This translates to the DAISY expression

$$
\begin{aligned}
&\text{FAC} : x0 \Longleftarrow \textbf{rec } \textit{Experiment}\\
&\quad\textbf{where}\\
&\qquad X =\ <x0\ !\ \text{DCR} : <X>>\\
&\qquad Y =\ <\ 1\ !\ \text{MPY} : <X\ Y>>\\
&\quad\text{READY} =\quad \text{ZERO?} : <X>\\
&\quad\text{VALUE} =\quad Y.
\end{aligned}
$$

Experiment is an expression stating what is to be observed about the circuit. Let us develop an experiment to display the entire circuit in operation. The obvious first attempt is the expression Print : $<X\ Y\ READY>$, but a display of this form would cause the signal X to be produced in its entirety. Hence, we would never get an opportunity to see Y and $READY$. The solution is to look at finite prefixes of each signal in turn. A transposed version of $<X\ Y\ READY>$ can be obtained by applying an identity component to the signal list. The interpreter transposes as a matter of course. We get a picture of the circuit in "time slices". Thus, the experiment we want is

$$< \lambda x.x\ *> \ : \ <X\ Y\ READY>$$

Let us generalize this experiment to work for other realizations. Define a function called Test that transposes any list of signals. In the figures that follow, a carriage return is interposed between time slices:

> Test : signal-list $\Leftarrow$ **rec** Format : Transpose : signal-list
> **where**
>> Transpose $= < \lambda x.x\ *>$
>> Format : [u ! v] $\Leftarrow$ $<$ *carriage-return* u ! Format : v $>$.

In the *factorial* example the desired experiment is

$$FAC\ :\ x \Leftarrow Test\ :\ <X\ Y\ READY>\ \textbf{where}....\ .$$

Figure 4.7 shows an interactive session in which this expression is executed[7]. As we fully expect, the first time interval that the X-register contains 0, the factorial of the initial value x^0 is found in the Y-register. In the next cycle the value is destroyed and X diminishes forever. The program must be interrupted to stop the display.

A similar experiment is run on the *FIB* realization in Figure 4.8. Again, the desired value appears as soon as the *READY* signal asserts its presence. It is worth noting that the circuit continues to compute valid Fibonacci numbers

---

[7]DAISY is implemented on a Digital Corporation VAX 11/780, under the UNIX operating system. Output from the DAISY sessions shown throughout this dissertation was recorded directly from the terminal by a host monitor program. These records have been modified as follows: some carriage-returns and blank characters are deleted; some blanks are replaced by tab characters to align columns. DAISY source listings are edited to the idealized syntax of Section 4.2; true source listings for each of the figures is shown in Appendix B. DAISY's prompt is an ampersand, '&'. The host interrupt character is EXT, typed control-C, and displayed as '↑C'.

afterward.

The *GCD* realization was

$$X = x^0 \ ! \quad U$$
$$Y = y^0 \ ! \quad \text{SUB(W, U)}$$
$$U = \qquad \text{MUX(V, X, Y)}$$
$$W = \qquad \text{MUX(V, Y, X)}$$
$$V = \qquad \text{LT?(X, Y)}$$
$$\text{READY} = \qquad \text{EQ?:(X, Y)}$$

Execution of the above is shown in Figure 4.9. Again the experiment is to trace the signals *X*, *Y*, and *READY* in parallel. This time, the circuit becomes completely stable two clock cycles after *READY* becomes true. Furthermore, the desired value of the computation is preserved—a desirable characteristic from the standpoint of integration—although its value ends up in the Y-register. The experiment has revealed a property of the realization that is not accounted for in its specification. The quality of becoming stable is in fact not expressible in the specification language as it now stands, since stability implies that the *GCD* circuit computes forever. The corresponding specification is an infinite loop whose formal meaning would be the totally undefined function. Thus, stability, like the quality of correctness, must be dealt with by some other means. We shall not develop this notion formally for it is not worth the effort. One can always impose stability on a circuit by using the *READY* signal to disable the registers. We return to the general issue of making assertions about performance in Chapter 7. For now, emulation of the circuit has at least given us a chance to discover an unspecified property without having either to build the circuit or to code it up in a simulation language. This is a significant practical advantage of this approach.

86

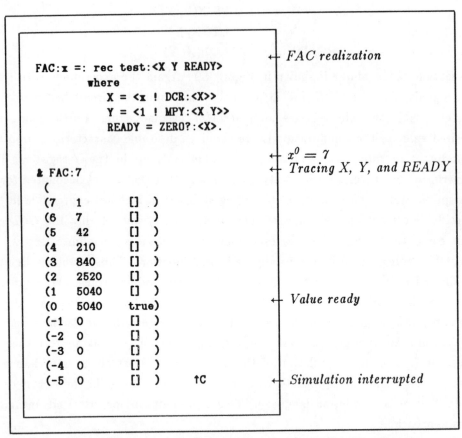

```
FAC:x =: rec test:<X Y READY>
        where
            X = <x ! DCR:<X>>
            Y = <1 ! MPY:<X Y>>
            READY = ZERO?:<X>.


& FAC:7
(
(7    1          []  )
(6    7          []  )
(5    42         []  )
(4    210        []  )
(3    840        []  )
(2    2520       []  )
(1    5040       []  )
(0    5040       true)
(-1   0          []  )
(-2   0          []  )
(-3   0          []  )
(-4   0          []  )
(-5   0          []  )      ↑C
```

← *FAC realization*

← $x^0 = 7$
← *Tracing X, Y, and READY*

← *Value ready*

← *Simulation interrupted*

Figure 4.7. Experiment with the FAC Realization.

```
FIB:x <= rec test:<X Y READY>
        where
              X = <x ! DCR:<X>>
              Y = <1 ! Z>
              Z = <1 ! ADD:<Y Z>>
              READY = ZERO?:<X>.


& FIB:7
 (
 (7      1          []  )
 (6      1          []  )
 (5      2          []  )
 (4      3          []  )
 (3      5          []  )
 (2      8          []  )
 (1      13         []  )
 (0      21         true)
 (-1     34         []  )
 (-2     55         []  )
 (-3     89         []  )
 (-4     144        []  )        ↑C
```

← *FIB realization*

← $x^0 = 7$
← *Tracing X, Y, and READY*

← *Value ready*

← *Simulation interrupted*

Figure 4.8. Experiment with the FIB Realization.

```
GCD:(x y) <= rec test:<X Y READY>        ←  GCD realization
            where
                X = <x ! U>
                Y = <y ! SUB:<W U>>
                U = IF:<V X Y>
                W = IF:<V Y X>
                V = LT?:<X Y>
                READY = EQ?:<X Y>


& GCD:(15 24)                            ←  x⁰ = 15, y⁰ = 24
   (                                     ←  Tracing X, Y, and READY
   (15   24      []   )
   (15   9       []   )
   (9    6       []   )
   (6    3       []   )
   (3    3       true)                   ←  Value ready
   (3    0       []   )
   (0    3       []   )
   (0    3       []   )
   (0    3       []   )
   (0    3       []   )        ↑C        ←  Simulation interrupted
```

Figure 4.9. Experiment with the GCD Realization.

5. Design Examples

We now have a language for describing digital circuits and a method to derive circuit descriptions from functional specifications. In this chapter, the method is applied to a larger example; a circuit is derived for a programming language interpreter. As descriptions get larger, it becomes necessary to organize them more carefully. We can "structure" circuit equations as we structure programs, by decomposing them hierarchically.

Since all of the structural combinations distribute over operator-lifting, we may arbitrarily *package* (*i.e.* give a name to) groups of interconnected combinatorial components. The instantaneous behavior of the packaged combination lifts to the signal behavior of the group.

We have already considered specifications that use complex data types, such as stacks. However, we have so far avoided building circuits over complex operators, by deriving equivalent specifications over more primitive types. In this chapter we finally face the task of implementing circuits over non-primitive signals. In programming, one hides implementation details by introducing *abstract data types*. We shall do the analogous thing at the behavioral level, introducing *abstract components* in our circuit descriptions. Like its programming counterpart, an abstract component is simply a specification of the external behavior required by the surrounding circuit.

Implementation of an object that has the right external behavior may be left as a subproblem. With the complex-typed signals factored out of the description, development of the controlling circuit can continue. As abstract components are factored from circuit descriptions, *instruction* signals are introduced to coordinate their behavior. Coordination of behavior forces us, for the first time, to consider the communicative qualities of the circuits we describe.

Hierarchical decomposition of large descriptions is common to all design realms. It is neither novel nor surprising that we do it with circuit descriptions, but is simply a necessary prelude to the attack of a larger design problem. Section 5.1 introduces some notation for structuring circuit descriptions. We exercise this notation on a small example that we have seen before. Since the example has to do with "language driven" design, we discuss that term in Section 5.2. In Section 5.3 we synthesize a circuit that interprets a programming language called L. The derivation is long and has five major steps. Recall that *synthesis* means a derivation that is not necessarily mechanizable. Indeed, there are numerous design decisions involved in the development of the L-circuit. We shall point out the transformations that require designer intervention as we present them.

The derivations that follow were done by hand. In Appendix B, the evolving specification is rewritten in DAISY. Executable versions of the specifications were quite helpful in debugging the derivation.

5.1. Higher Level Components

In Section 2.4 we used the Wand-Friedman transformation strategy to synthesize a stacking version of the non-linear scheme

$$F(x) \Leftarrow p(x) \rightarrow f(x),\ h(\ F(g_0(x)),\ F(g_1(x))\ ).$$

We arrived at the form

$$G(x, \sigma) \Leftarrow p(x) \rightarrow R(f(x), \sigma),\ G(g_0(x),\ push(0,\ push(g_1(x),\ \sigma)))\ ).$$

$$R(v, \sigma) \Leftarrow empty?(\sigma) \rightarrow v,$$
$$eq?(top(\sigma),\ 0) \rightarrow G(top(pop(\sigma)),\ push(1,\ push(v,\ pop(pop(\sigma)))\ )\ ),$$
$$R(\ h(top(pop(\sigma)),\ v),\ pop(pop(\sigma))\ ).$$

For the purposes of this discussion, we shall separate the recursion stack into two stacks: τ holds actions and σ holds values. Since every recursive call pushes exactly one action and one value, the modification is trivial. In addition, let us use truth values $\{tt,\ ff\}$ to denote actions, there being only two. The revised specification is

$$G(x, \sigma, \tau) \Longleftarrow p(x) \rightarrow R(f(x), \sigma, \tau),\ G(g_0(x), push(g_1(x), \sigma)),\ push(\mathit{tt}, \tau)\ ).$$

$$R(v, \sigma, \tau) \Longleftarrow empty?(\tau) \rightarrow v,$$
$$top(\tau) \rightarrow G(top(\sigma), push(v, pop(\sigma)), push(\mathit{ff}, pop(\tau))\ ),$$
$$R(\ h(top(\sigma), v),\ pop(\sigma),\ pop(\tau)\ ).$$

By the construction of Section 2.4.4 this specification is transformed to a system with a single function variable symbol. A new control token, w, encodes which of G or R is in control. The identifier, v, in the definition of the function R is changed to x in order to give the system a uniform formal argument list.

$$H(w, x, \sigma, \tau) \Longleftarrow$$
$$at?(w, \mathbf{G}) \rightarrow [p(x) \rightarrow H(\mathbf{R}, f(x), \sigma, \tau),\ H(\mathbf{G}, g_0(x), push(\mathit{tt}, \tau), push(g_1(x), \sigma))\ ],$$
$$at?(w, \mathbf{R}) \rightarrow [empty?(\tau) \rightarrow x,$$
$$top(\tau) \rightarrow H(\mathbf{G}, top(\sigma), push(\mathit{ff}, pop(\tau)), push(x, pop(\sigma))),$$
$$H(\mathbf{R}, h(\ top(\sigma),\ x),\ pop(\tau),\ pop(\sigma))\ ].$$

By distributing the conditional, we can turn this equation into an instance of U_I. After a little algebra on the resulting terms we arrive at the equation

$$H(w, x, \sigma, \tau) \Longleftarrow and(at?(w, \mathbf{R})\ empty?(\tau)) \rightarrow x,$$
$$H(\ mux(at?(w, \mathbf{G}),\ mux(p(x), \mathbf{R}, \mathbf{G}),\ mux(top(\tau), \mathbf{G}, \mathbf{R})\ ),$$
$$mux(at?(w, \mathbf{G}),\ mux(p(x), 1, g_0(x)),\ mux(top(\tau), h(top(\sigma), x), top(\sigma)\ )),$$
$$mux(at?(w, \mathbf{G}),\ mux(p(x), \sigma,\ push(g_1(x), \sigma),$$
$$mux(top(\tau), pop(\sigma),\ push(x, pop(\sigma))\ )),$$
$$mux(at?(w, \mathbf{G}),\ mux(p(x), \tau,\ push(\mathit{ff}, \tau),\ mux(top(\tau),\ pop(\tau),\ push(\mathit{tt}, pop(\tau))\ )),$$

We shall adapt some familiar structured programming techniques to decompose the realization of this function.

5.1.1. Packaged Combinations.

Let us introduce a more sophisticated multiplexor to take advantage of the fact that the conditional structure of each inner call is the same. Define a combined operator that does four-way selection.

$$mux_4(p,\ q,\ r,\ u,\ v,\ w,\ x) \Longleftarrow mux(p,\ mux(q,\ u,\ v),\ mux(r,\ w,\ x)\ ).$$

We should perhaps call the combination something like "3-by-4 selector"; the name mux_4 is used for brevity. Using mux_4 we can rewrite H's defining equation as

$$H(w,\ x,\ \sigma,\ \tau) \Longleftarrow and(at?(w,\ R)\ empty?(\tau)) \to x,$$
$$H(\ mux_4(at?(w,\ G),\ p(x),\ top(\tau),\ R,\ G,\ G,\ R),$$
$$mux_4(at?(w,\ G),\ p(x),\ top(\tau),\ f(x),\ g_0(x),\ h(top(\sigma),\ x),\ top(\sigma)\ ),$$
$$mux_4(at?(w,\ G),\ p(x),\ top(\tau),\ \sigma,\ push(g_1(x),\ \sigma),\ pop(\sigma),\ push(x,\ pop(\sigma))\ ),$$
$$mux_4(at?(w,\ G),\ p(x),\ top(\tau),\ \tau,\ push(\mathit{ff},\ \tau),\ pop(\tau),\ push(\mathit{tt},\ pop(\tau))\ )).$$

It is not an accident that mux_4 fails to absorb all the shared subexpressions. The reason is evident when H is transcribed to a circuit description. As before, lifted operations are written in upper case. Lifted constants are enclosed in square braces '[' and ']' to distinguish them from signals and components.

$$C(\ w^0,\ x^0,\ \sigma^0,\ \tau^0\ ) \Longleftarrow$$

```
W = w^0 ! MUX_4(U, V, Y, [R], [G], [R], [G])
X = x^0 ! MUX_4(U, V, Y, F(X), G_0(X), H(TOP(Σ), X), TOP(Σ))
Σ = σ^0 ! MUX_4(U, V, Y, Σ, PUSH(G_1(X), Σ), POP(Σ), PUSH(X, POP(Σ)) )
T = τ^0 ! MUX_4(U, V, Y, T, PUSH([ff], T), POP(T), PUSH([tt], POP(T)) ).
U =        AT?(W, [G])
V =        P(X)
Y =        TOP(T)
READY =    AND(AT?(W, [R]), EMPTY?(T))
VALUE =    X
```

The outputs of the components $AT?$, P, and TOP are shared by all instances of MUX_4. Had the subexpressions $at?(l,\ G)$, $p(x)$, and $top(\tau)$ been incorporated in the definition of mux_4, each instance of the multiplexor would have included a duplicate set of the predicate components. While duplication is not necessarily a bad thing, we elect to avoid it here. Combined operation mux_4 can be lifted to component MUX_4 because mux_4 is defined by a trivial expression. By Propositions 3.3-1 through 3.3-4, the combination is transparent to lifting.

5.1.2. Abstract Components. While circuit C certainly computes the same thing as H, and hence as the original specification F, it is hard to justify calling it a realization. Its registers Σ and T range over stacks, and so there is much yet to do before going to the laboratory with this circuit description. We should think of stacks abstractly and hide their implementation details. Let us therefore introduce a "class object" that gives the necessary information about a stack: what its top is and whether the stack is empty. That is, we shall replace stack *objects*, to which operations are applied directly, with stack *agents*, which can be instructed to apply those operations. Separate the signals that have to do with the two stacks, and rewrite the realization as

$$
\begin{aligned}
\mathrm{C}(w^0,\ x^0,\ \sigma^0,\ \tau^0\ ) &\Longleftarrow \\
\mathrm{W} = w^0\ &!\ \ \mathrm{MUX}_4(\mathrm{U},\ \mathrm{V},\ \mathrm{Z_T},\ [\mathrm{R}],\ [\mathrm{G}],\ [\mathrm{R}],\ [\mathrm{G}]) \\
\mathrm{X} = x^0\ &!\ \ \mathrm{MUX}_4(\mathrm{U},\ \mathrm{V},\ \mathrm{Z_T},\ \mathrm{F}(\mathrm{X}),\ \mathrm{G}_0(\mathrm{X}),\ \mathrm{H}(\mathrm{Z}_\Sigma,\ \mathrm{X}),\ \mathrm{Z}_\Sigma) \\
\mathrm{U} =\ &\quad \mathrm{AT?}(\mathrm{W},\ [\mathrm{G}]) \\
\mathrm{V} =\ &\quad \mathrm{P}(\mathrm{X}) \\
\mathrm{Z} =\ &\quad \mathrm{G}_1(\mathrm{X}) \\
\mathrm{READY} =\ &\quad \mathrm{AND}(\mathrm{AT?}(\mathrm{W},\ [\mathrm{R}]),\ \mathrm{E_T}) \\
\mathrm{VALUE} =\ &\quad \mathrm{X} \\[6pt]
\Sigma = \sigma^0\ &!\ \ \mathrm{MUX}_4(\mathrm{U},\ \mathrm{V},\ \mathrm{Z_T},\ \Sigma,\ \mathrm{PUSH}(\mathrm{Z},\ \Sigma),\ \mathrm{POP}(\Sigma),\ \mathrm{PUSH}(\mathrm{X},\ \mathrm{POP}(\Sigma))\ ) \\
\mathrm{E}_\Sigma =\ &\quad \mathrm{EMPTY?}(\Sigma) \\
\mathrm{Z}_\Sigma =\ &\quad \mathrm{TOP}(\Sigma) \\[6pt]
\mathrm{T} = \tau^0\ &!\ \ \mathrm{MUX}_4(\mathrm{U},\ \mathrm{V},\ \mathrm{Z_T},\ \mathrm{T},\ \mathrm{PUSH}([\mathit{ff}],\ \mathrm{T}),\ \mathrm{POP}(\mathrm{T}),\ \mathrm{PUSH}([\mathit{tt}],\ \mathrm{POP}(\mathrm{T}))\ ). \\
\mathrm{E_T} =\ &\quad \mathrm{EMPTY?}(\mathrm{T}) \\
\mathrm{Z_T} =\ &\quad \mathrm{TOP}(\mathrm{T})
\end{aligned}
$$

Two signals have been added and one name has been changed, in order to bring out the similarity between the stacking subcircuits. Y is renamed E_T, the action-stacke's "I-am-empty" signal. The corresponding signal E_Σ for the value-stack is not used but is included in the description for symmetry. The new signal identifier Z was introduced because the ability to do the operation G_1 should not be ascribed to the behavior of the stack.

Our next goal is to hide all the pushing and popping inside of a component definition. We must only ensure that the new component's external behavior, the values on the signals Z_Σ, E_Σ, Z_T, and E_T, is the same as before. As they stand however, the equations that specify these behaviors are too specific, for they inherit their decision making apparatus from C. The stack agents must be able to push, pop, replace the top of, or do nothing with the stacks in their care. It should be left to the surrounding circuit to determine which of these operations to perform. Introduce a set of *instructions*, $Inst = \{\textbf{NOOP}, \textbf{PUSH}, \textbf{POP}, \textbf{PLOP}\}$, and define a component

$$STACK : (Stack \times Sig_{Ins} \times Sig_{Val}) \rightarrow (Sig_{Val} \times Sig_{Bool})$$

that makes the instructions work.

$$STACK(\sigma^0, INSTRUCTION, VALUE) \Longleftarrow$$

    rec

      $(TOP(\Sigma), EMPTY?(\Sigma))$

    where

      $\Sigma = \sigma^0 \,!\, operate^\infty(INSTRUCTION, VALUE, \Sigma)$

      $operate(instruction, value, stack) \Longleftarrow$

      $eq?(instruction, \textbf{NOOP}) \rightarrow \sigma,$

        $eq?(instruction, \textbf{POP}) \rightarrow pop(\sigma),$

        $eq?(instruction, \textbf{PUSH}) \rightarrow push(value, \sigma),$

        $eq?(instruction, \textbf{PLOP}) \rightarrow push(value, pop(\sigma)).$

Now if C can be made to generate the right instructions at the right times, *STACKs* can be used in place of the signals Σ and T. Determination of the appropriate instructions is easy; it is given by the original signal definitions in C. The *STACKs* for Σ and T can share an instruction signal, I.

$$C(\ w^0,\ x^0,\ \sigma^0,\ \tau^0\ ) \Leftarrow$$

$$W = w^0\ !\ MUX_4(U,\ V,\ Z_T,\ [\mathbf{R}],\ [\mathbf{G}],\ [\mathbf{R}],\ [\mathbf{G}])$$

$$X = x^0\ !\ MUX_4(U,\ V,\ Z_T,\ F(X),\ G_0(X),\ H(Z_\Sigma,\ X),\ Z_\Sigma)$$

$$U = \quad AT?(W,\ [\mathbf{G}])$$

$$V = \quad P(X)$$

$$Z = \quad G_1(X)$$

$$READY = \quad AND(AT?(W,\ [\mathbf{R}]),\ E_T)$$

$$VALUE = \quad X$$

$$I = \quad MUX_4(U,\ V,\ Z_T,\ [\mathbf{NOOP}],\ [\mathbf{PUSH}],\ [\mathbf{POP}],\ [\mathbf{PLOP}])$$

$$(Z_\Sigma,\ E_\Sigma) = \quad STACK(\ \sigma^0,\ I,\ MUX_4(U,\ V,\ Z_T,\ \phi,\ Z,\ \phi,\ X\ ))$$

$$(Z_T,\ E_T) = \quad STACK(\ \tau^0,\ I,\ MUX_4(U,\ V,\ Z_T,\ \phi,\ [\![f]\!],\ \phi,\ [\![t]\!]\ ))$$

The circuit has been factored into abstract components that communicate with instructions. The factorization is an application of conditional distributivity to operations (Sec. 2.6.3). In more detail, the "next" value for the stack σ is an expression of the form

$$p \rightarrow [q \rightarrow \sigma,\ push(u,\ \sigma)],\ [r \rightarrow pop(\sigma),\ push(v,\ pop(\sigma))]$$

where p, q, and r are the appropriate propositional terms. Let us "normalize" the operations to make way for the factorization. Introduce combined operations

$$noop'\ (x,\ \sigma) \Leftarrow \sigma. \qquad push'\ (x,\ \sigma) \Leftarrow push(x,\ \sigma).$$
$$pop'\ (x,\ \sigma) \Leftarrow pop(\sigma). \qquad plop'\ (x,\ \sigma) \Leftarrow push(x,\ pop(\sigma)).$$

The conditionals distribute over operations and operands alike.

$$apply(\ [p \rightarrow [q \rightarrow noop',\ push'\ ],\ [r \rightarrow pop',\ plop'\ ]]\ ,$$
$$[p \rightarrow [q \rightarrow \phi,\ u],\ [r \rightarrow \phi,\ v]]\ ,$$
$$[p \rightarrow [q \rightarrow \sigma,\ \sigma],\ [r \rightarrow \sigma,\ \sigma]]\ )$$

To lift this expression we need to think in terms of a component *APPLY* whose inputs include the signal

$$MUX_4(P,\ Q,\ R,\ [\ noop\ ],\ [\ push\ ],\ [\ pop\ ],\ [\ plop\ ]\ )$$

However, it is counterintuitive to assert that operations are legitimate values for

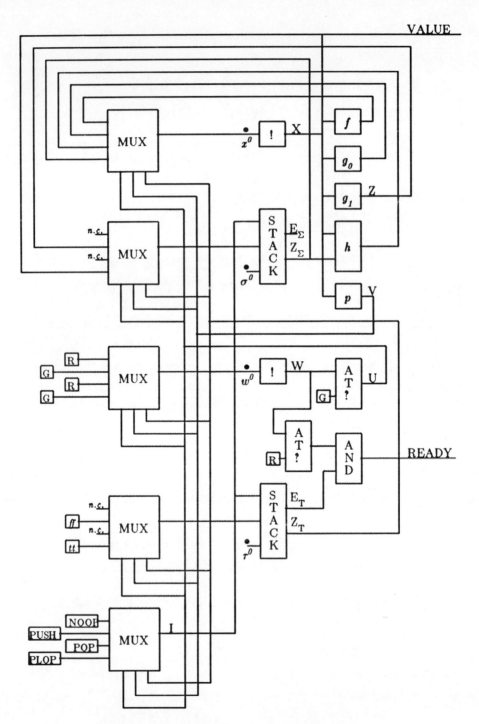

Figure 5.1. A Schematic for Circuit C.

a signal to hold. The physical interpretation must be that the selected operation is encoded as an instruction to be interpreted by the abstracted subcircuit. Essentially the same principle is involved when we introduce a control token. It is this technique of factorization that motivated the decision to model a component as a signal in Section 3.5.

Figure 5.1 gives a schematic version of the circuit description for C. Since we began with a recursion scheme the realization is a generalization, with components f, g_0, g_1, h, and p being variable. The *Fibonacci* function is an instance of the original non-linear specification

$$F(x) \iff p(x) \to f(x),\ h(\ F(g_0(x)),\ F(g_1(x))\ ).$$

with

$$p(u) \iff lt?(u,\ 2).$$

$$f(u) \iff 1.$$

$$h(u,\ v) \iff add(u,\ v).$$

$$g_0(u) \iff dcr(dcr(u)).$$

$$g_1(u) \iff dcr(u).$$

The corresponding instance of circuit C, in which these packaged combinations replace the component variables, realizes *FIB* provided it halts. (Recall that the stack transformation may have weakened the resulting specification.) The controlling circuit for specification *FIB* is

$$C_{FIB}(\ w^0,\ x^0,\ \sigma^0,\ \tau^0\ ) \Longleftarrow$$

$$W = w^0\ !\ \text{MUX}_4(\text{U, V, Z}_\text{T},\ [\text{R}],\ [\text{G}],\ [\text{R}],\ [\text{G}])$$

$$X = x^0\ !\ \text{MUX}_4(\text{U, V, Z}_\text{T},\ 1,\ \text{DCR(DCR(X))},\ \text{ADD(X, Z}_\Sigma),\ \text{Z}_\Sigma)$$

$$U = \quad \text{AT?(W,}\ [\text{G}])$$

$$V = \quad \text{LT?(X, 2)}$$

$$Z = \quad \text{DCR(X)}$$

$$\text{READY} = \quad \text{AND(AT?(W,}\ [\text{R}]),\ \text{E}_\text{T})$$

$$\text{VALUE} = \quad X$$

$$I = \quad \text{MUX}_4(\text{U, V, Z}_\text{T},\ [\text{NOOP}],\ [\text{PUSH}],\ [\text{POP}],\ [\text{PLOP}])$$

$$(\text{Z}_\Sigma,\ \text{E}_\Sigma) = \quad \text{STACK}(\ \sigma^0,\ \text{I, MUX}_4(\text{U, V, Z}_\text{T},\ \phi,\ \text{Z},\ \phi,\ \text{X}))$$

$$(\text{Z}_\text{T},\ \text{E}_\text{T}) = \quad \text{STACK}(\ \tau^0,\ \text{I, MUX}_4(\text{U, V, Z}_\text{T},\ \phi,\ [\textit{ff}],\ \phi,\ [\textit{tt}]))$$

Figure 5.2 shows the usual DAISY experiment on C_{FIB} with stacks implemented as lists (see Appendix B). We have introduced techniques to structure circuit descriptions by decomposing them into hierarchies of *higher level components*. Packaged combinations such as MUX_4 serve as macros that identify repeatedly used connection patterns. Their introduction is valid because operator combination is transparent to lifting. Abstract components are the behavioral analog of Hoare's abstract data types (1972). To hide implementation details, signals over complex values are replaced by agents that manage those values. The factorization introduces instructions generated by the surrounding circuit. While we have not provided a plausible realization for *STACK* components, we have succeeded in isolating the task and can proceed with the refinement of the controlling circuit.

Deciding how much of the surrounding circuit to incorporate into a higher level component is non-trivial. Had MUX_4 included predicates P and $AT?$, they would have been duplicated in every instance of MUX_4, and the opportunity to share some of the computation would have been lost. Had the Σ-*STACK* description retained its ability to compute G_1 it would have been too specialized to reveal its similarity to the *T-STACK*.

```
FIBckt:(w0 x0 s0 t0) <= rec test:<READY X I V1 W V2 E2 U V>
   where
         W = <w0 ! MUX-N:<P Q V2 [R*] [G*] [R*] [G*]> >
         X = <x0 ! MUX-N:<P Q V2 [1*] DCR:<DCR:<X>> ADD:<X V1> V1> >
    [V1 E1] =         STACK:<s0 I MUX-N:<P Q V2 [??*] DCR:<X> [??*] X>>
    [V2 E2] =         STACK:<t0 I MUX-N:<P Q V2 [??*] [<>*] [??*] [tt*]>>
         I =          MUX-N:<P Q V2 [noop*] [push*] [pop*] [plop*]>
         U =          AT?:<W [G*]>
         V =          LT?:<X [2*]>
     READY =          AND:<AT?:<W [R*]> E2>.


MUX-N = [mux-N*].
mux-N:[p q r u v w x] <= mux:<p mux:<q u v> mux:<r w x>>.


STACK:[s0 I V] <= rec <<top*>:<S> <empty?*>:<S>>
   where
       S = <s0 ! <operate*>:<I V S>>
       operate:[i v s] <=
          same?:<i @noop> -> s,
          same?:<i @pop > -> pop:s,
          same?:<i @push> -> push:<v s>,
          same?:<i @plop> -> plop:<v s>>.
```

Figure 5.2a. Experiment with C_{FIB} – Source for the Realization.

(See Appendix B for the implementation of stacks.)

```
fib:n <= FIBckt:<0 n MTstk MTstk>.                    ←  Register setup

& fib:4                                               ←  Find FIB(4)
(                                                     ←  Tracing READY,
  ([]    4   push   ??  0  ??   true    true   []  )     X, I, V₁, W,
  ([]    2   push   3   0  []   []      true   []  )     V₂, E₂, U, V
  ([]    0   noop   1   0  []   []      true   true)  (See Figure 5.2a)
  ([]    1   plop   1   1  []   []      []     true)
  ([]    1   noop   1   0  tt   []      true   true)
  ([]    1   pop    1   1  tt   []      []     true)
  ([]    2   plop   3   1  []   []      []     []  )
  ([]    3   push   2   0  tt   []      true   []  )
  ([]    1   noop   2   0  []   []      true   true)
  ([]    1   plop   2   1  []   []      []     true)
  ([]    2   push   1   0  tt   []      true   []  )
  ([]    0   noop   1   0  []   []      true   true)
  ([]    1   plop   1   1  []   []      []     true)
  ([]    1   noop   1   0  tt   []      true   true)
  ([]    1   pop    1   1  tt   []      []     true)
  ([]    2   pop    1   1  tt   []      []     []  )
  ([]    3   pop    2   1  tt   []      []     []  )
  (true  5   pop    ??  1  ??   true    []     []  )

  (true 73404895/14680979 pop ?? 1 ?? true [] [])    ←  Value ready
  (true 73404895/14680979 pop ?? 1 ?? true [] [])
  (true 73404895/14680979 pop ?? 1 ?? true [] [])    ←  Value lost.
  (true 73404895/14680979 pop ?? 1 ?? true [] [])
  (true 73404895/14680979 pop ?? 1 ?? true [] [])
                                                     ←  Simulation
                                                        interrupted
```

Figure 5.2b. Experiment with C_{FIB} – Record of an Experiment.

5.2. Language Driven Design – Introduction

Let us briefly consider a different instance of the realization C, derived in the previous section. The same circuit description scheme gives an evaluator for arithmetic expressions, specified by a semantic function similar to the one in Section 2.6.1. The argument x will range over expressions in a language *Exp*

$$expression ::= atom \mid (\ expression + expression\ )$$

Assume operations *left : Exp $\rightarrow$ Exp* and *right : Exp $\rightarrow$ Exp* that return left and right subexpressions; *atom? : Exp $\rightarrow$ Bool* that distinguishes atomic expressions; *fetch? : Atom $\rightarrow$ Int* that produces numbers from atoms; and *opn : (Int $\times$ Int) $\rightarrow$ Int*, an arithmetic operation. The recursion equation

$$\mathbb{E}(x) \quad \Longleftarrow \quad atom?(x) \rightarrow fetch(x),\ opn(\ \mathbb{E}(left(x)),\ \mathbb{E}(right(x))\ ).$$

defines the value of any expression. Since $\mathbb{E}$'s defining equation is an instance of the non-linear specification of the preceding section, the corresponding instance of C realizes $\mathbb{E}$.

$$
\begin{array}{ll}
\multicolumn{2}{l}{\mathrm{C}(w^0,\ x^0,\ \sigma^0,\ \tau^0\ ) \Longleftarrow} \\[4pt]
\mathrm{W} = w^0\ ! & \mathrm{MUX}_4(\mathrm{U},\ \mathrm{V},\ \mathrm{Z_T},\ [\mathbf{R}],\ [\mathbf{G}],\ [\mathbf{R}],\ [\mathbf{G}]) \\[2pt]
\mathrm{X} = x^0\ ! & \mathrm{MUX}_4(\mathrm{U},\ \mathrm{V},\ \mathrm{Z_T},\ \mathrm{FETCH(X)},\ \mathrm{LEFT(X)},\ \mathrm{OPN}(\mathrm{Z_\Sigma},\ \mathrm{X}),\ \mathrm{Z_\Sigma}) \\[2pt]
\mathrm{U} = & \mathrm{AT?(W,}\ [\mathbf{G}]) \\[2pt]
\mathrm{V} = & \mathrm{ATOM?(X)} \\[2pt]
\mathrm{Z} = & \mathrm{RIGHT(X)} \\[2pt]
\mathrm{READY} = & \mathrm{AND(AT?(W,}\ [\mathbf{R}]),\ \mathrm{E_T}) \\[2pt]
\mathrm{VALUE} = & \mathrm{X} \\[6pt]
\hline \\[-6pt]
\mathrm{I} = & \mathrm{MUX}_4(\mathrm{U},\ \mathrm{V},\ \mathrm{Z_T},\ [\mathbf{NOOP}],\ [\mathbf{PUSH}],\ [\mathbf{POP}],\ [\mathbf{PLOP}]) \\[2pt]
(\mathrm{Z_\Sigma},\ \mathrm{E_\Sigma}) = & \mathrm{STACK}(\sigma^0,\ \mathrm{I},\ \mathrm{MUX}_4(\mathrm{U},\ \mathrm{V},\ \mathrm{Z_T},\ \phi,\ \mathrm{Z},\ \phi,\ \mathrm{X}\ )) \\[2pt]
(\mathrm{Z_T},\ \mathrm{E_T}) = & \mathrm{STACK}(\tau^0,\ \mathrm{I},\ \mathrm{MUX}_4(\mathrm{U},\ \mathrm{V},\ \mathrm{Z_T},\ \phi,\ [f\!f],\ \phi,\ [t\!t]\ ))
\end{array}
$$

The circuit is a "direct interpreter" for a suitably represented language of arithmetic expressions. It calculates a value by processing the expression itself, saving both intermediate results and subexpressions on its stack. Non-atomic expressions are evaluated left-to-right, since that was the order imposed by the stacking transformation. A variety of improvements in the design are possible, of course. We might arrange some form of look-ahead to keep from stacking some atomic subexpressions. This refinement can be developed formally by first unfolding E to expose more tests:

$$E(e) \Leftarrow atom?(e) \rightarrow fetch(e),$$
$$atom?(left(e)) \rightarrow opn(\ fetch(left(e)),\ E(right(e))\ ),$$
$$atom?(right(e)) \rightarrow opn(\ E(left(e)),\ fetch(right(e))\ ),$$
$$opn(\ E(left(e)),\ E(right(e))\ ).$$

and then transforming to circuit form.

A more conventional architecture would not stack text at all, but requires a compiler to translate expressions into sequential programs. Wand (1982a, 1982b) develops a method for deriving compiler/machine pairs that yield more classic stored program organizations. His derivations lead to iterative machine specifications and can therefore be immediately extended to obtain circuit descriptions of the machines.

5.3. Application to Language Driven Design

The derivation techniques we have developed so far are used below to synthesize a realization from a programming language specification. The target circuit is a direct interpreter for expressions in an applicative language called L. The derivation has six major steps. All but the first are transformations; of the five transformations, two are direct constructions. To varying degrees, the remaining steps involve designer creativity, and thus are at best semi-mechanizable.

We begin with a formal definition of L's semantics. This fully abstract specification is then rewritten as a function on represented expressions. Hence, the first step is to turn L's formal definition into something concrete enough to be regarded as a program, an *L-interpretor*. Readers uncomfortable with the mathematics can skim the details on first reading, and take the interpreter specification (Figure 5.4) as the starting point for synthesis.

The initial specification is non-linear. The second derivation step introduces a recursion stack to linearize control. As we have mentioned before, this is regarded as a creative step because recursive calls must be ordered. The resulting interpreter implements an applicative order computation rule for L and is, therefore, only partially equivalent to the initial specification.

The stacking version of the interpreter is compiled into a loop by introducing a control token. The result could be transformed to a circuit; however, some refinements are made that lead to a more compact specification. These refinements depend on subtle representation issues and this derivation step depends more than any of the others on intelligent guidance.

The refined loop is then transcribed to a realization, and the last step in the derivation factors the system into abstract components.

Like most lengthy presentations, this one tells little of what motivated specific design decisions. The product of the synthesis is described, without discussion of the blind alleys, or the discovery of features that reflected negatively on earlier specifications. At each step of the derivation a version of the evolving specification was written in DAISY. Experimentation revealed flaws in some design refinements, and a number of typographical errors. The DAISY versions, and some trial experiments, are shown in Appendix B.

5.3.1. The language L. *L* is a purely applicative, lexically scoped language with constructs for programmer-defined functions and self-referential values. Its formal definition is given in Figure 5.3. All operators and defined functions are 1-placed. One writes "(add : n) : m" to add two numerals; the operation *add* returns a second operation that "adds *n* " (Parentheses show how expressions should be parsed). Assume that the operator set includes {*zero?*, *one?*, *inc*, *dcr*, *lt?*, *eq?*, *add*, *sub*, *mpy* }. The operations associated with these names are held in an initial environment.

Label-expressions[1] are used to define functions recursively. Our three

---

[1] The form "*i* ⇐ *e* " is analogous to the LISP expression "(LABEL I E)" (McCarthy, *et.al.*, 1965). While any expression may occur to the right of the "assignment" symbol, it is not immediately clear what expressions are sensible there. For example the form "x ⇐ inc:x" does *not* have the effect of setting *x* to *x* +1, but instead diverges. That λ-expressions are meaningful in label-expressions depends in part on the fact that they evaluate to *closures*, that is, data structures that incorporate environmental information. (McCarthy's LABEL requires *e* to be a LAMBDA-expression.) To allow other non-trivial forms, we need primitive operators that return closures. A suspending *CONS* would do nicely.

Expression Syntax

$$Exp \quad Ide \mid Nml \mid \lambda \, i \, . \, e \mid i \Leftarrow e \mid e_1 : e_2 \mid e_1 \rightarrow e_2 \, , \, e_3$$

Domains

Ide	*(i)*	*identifiers*
Num	*(n)*	*numerals*
Bool	*(b)*	*truth values*
$Opn = Val \rightarrow Bas$	*(o)*	*operations*
$Err = \{\,\text{``invalid function''},...\}$	*(m)*	*error messages*
Exp	*(e)*	*expressions*
$Bas = Num + Bool + Opn + Err$	*(v)*	*basic values*
$Val = Bas + Ftn$	*(v)*	*expressable values*
$Ftn = Val \rightarrow Val$	*(f)*	*functions*
$Env = Ide \rightarrow Val$	*(ρ)*	*environments*

Valuation $L : Exp \rightarrow Env \rightarrow Val$

$$L \llbracket\, n \,\rrbracket \rho = n$$
$$L \llbracket\, i \,\rrbracket \rho = \rho(i)$$
$$L \llbracket\, \lambda \, i \, . \, e \,\rrbracket \rho = \lambda v. \; L \llbracket\, e \,\rrbracket \, (\rho[v/i])$$
$$L \llbracket\, i \Leftarrow e \,\rrbracket \rho = fix \, (\, \lambda \epsilon. \; L \llbracket\, e \,\rrbracket \, \rho[\epsilon/i])$$
$$L \llbracket\, e_1 : e_2 \,\rrbracket \rho = apply \, (L \llbracket\, e_1 \,\rrbracket \rho) \, (L \llbracket\, e_2 \,\rrbracket \rho)$$
$$L \llbracket\, e_1 \rightarrow e_2 \, , \, e_3 \,\rrbracket \rho = test(L \llbracket\, e_1 \,\rrbracket \rho) \rightarrow L \llbracket\, e_2 \,\rrbracket \rho, \; L \llbracket\, e_3 \,\rrbracket \rho$$

Auxiliaries

$$\rho[v/i] = \lambda x. \; (x = i) \rightarrow v, \; \rho(x).$$
$$apply = \lambda fv. \; (f \, isOpr) \rightarrow fv, \; (f \, isFtn) \rightarrow fv, \; \text{``invalid function''}.$$
$$test = \lambda v. \; (v \, isBool) \rightarrow (v \, asBool), \; f\!f.$$

Figure 5.3. Standard Semantics of the Language L.

example functions are expressed as follows in L:

$$\text{GCD} \Leftarrow \lambda\, x\, .\, \lambda\, y\, .$$
$$(\text{eq?:x}) : y \to x\, ,$$
$$(\text{lt?:x}) : y \to (\text{GCD:x}) : ((\text{sub:y):x}),\ (\text{GCD:y}) : ((\text{sub:x):y})$$

$$\text{FAC} \Leftarrow \lambda\, x\, .\ \text{zero?:x} \to 1,\ (\text{mpy:x}) : (\text{FAC:(dcr:x)})$$

$$\text{FIB} \Leftarrow \lambda\, x\, .\ (\text{lt?:x}) : 2 \to 1,\ (\text{add:(FIB:(dcr:x)))} : (\text{FIB:(dcr:(dcr:x))})$$

These forms are used for the benchmark tests in Appendix B.

5.3.2. An L-Interpreter.

We follow Wand's advice to compiler designers (1982a). Given the semantic function $\mathbb{L} : Exp \to Env \to Val$, we seek a machine description, $\mathbb{M}$, corresponding to $\mathbb{L}$. However, while $\mathbb{L}$ acts on abstract entities, $\mathbb{M}$ acts on their *representations*. Some notation is helpful. Given a domain, D, let Rep_D denote a *representation of D*. If α is in Rep_D, denote the thing α represents by $\overset{\triangledown}{\alpha}$. When α is a complex expression, we shall write $^\triangledown[\alpha]$.

One of the tasks of a compiler is to produce program representations from expressed programs. The machine must interpret compiled programs consistently. That is, given a compiler $\mathbb{R} : Exp \to Rep_{Exp}$, and a machine

$$\mathbb{M} : (Rep_{Exp} \times Rep_{Env}) \to Rep_{Val},$$

we require that

$$^\triangledown[\mathbb{M}(\mathbb{R}(exp),\ env)] = \mathbb{L} [\![\ exp\ ]\!] \overset{\triangledown}{env}$$

Since we are deriving a direct interpreter for L, $\mathbb{R}$ is a trivial translator, and we omit reference to it by asking instead that

$$^\triangledown[\mathbb{M}(exp,\ env)] = \mathbb{L} [\![\ \overset{\triangledown}{exp}\ ]\!] \overset{\triangledown}{env}.$$

We will assemble $\mathbb{M}$'s specification by attempting to rewrite $\mathbb{L}$ as the analogous function on concrete representations. Along the way, new objects will be discovered that require representational counterparts, and some of the properties of these objects will have to be inherited by their representations. Which properties to preserve are revealed when we try to prove $\mathbb{M}$'s correctness.

Representations are expressed as records delimited with square brackets, '[' and ']'. Within the delimiters are a sequence of field names, the first of which is always a *tag*. For example, represented expressions (discussed just below) have record structure [*tag lft rgt*]. With each record format there are predicates, field extractors, and record builders, designated by the associated field name. For example, expressions have field extractors *tag, lft,* and *rgt.* Since NUM is a possible expression tag, there is a predicate *num?* that tests for that tag, and a constructor *make*-NUM that builds numeric expressions.

Expression Representation.

Of the six kinds of expressions, only the conditional has more than two subexpressions. Let a represented expression be a record of three fields, [*tag lft rgt*], where *tag* is one of {NUM, IDE, LAM, LBL, APL, CND, TST}. Define the translator $I\!R$ as follows

$$I\!R [\![n]\!] = [\text{NUM } n \, \phi]$$

$$I\!R [\![i]\!] = [\text{IDE } i \, \phi]$$

$$I\!R [\![\lambda \, i \, . \, e]\!] = [\text{LAM } i \; I\!R [\![e]\!] \,]$$

$$I\!R [\![i \Leftarrow e]\!] = [\text{LBL } i \; I\!R [\![e]\!] \,]$$

$$I\!R [\![e_1 : e_2]\!] = [\text{APL } I\!R [\![e_1]\!] \; I\!R [\![e_2]\!] \,]$$

$$I\!R [\![e_1 \to e_2 , e_3]\!] = [\text{CND } I\!R [\![e_1]\!] \; [\text{TST } I\!R [\![e_2]\!] \; I\!R [\![e_3]\!] \,] \,]$$

We suppress unused fields, and write "[IDE i]" rather than "[IDE i ϕ]".

Environment Representation.

We shall not define a detailed record structure for Rep_{Env}. Instead, just assume that operations

$$\text{find} : (Ide \times Rep_{Env}) \to Rep_{Val}$$
$$\text{extend} : (Rep_{Env} \times Rep_{Val} \times Ide) \to Rep_{Env}$$

exist that satisfy

$$\triangledown[\text{find}(i, env)] = \overset{\triangledown}{env}(i)$$

$$\triangledown[\text{extend}(env, val, i)] = \overset{\triangledown}{env} [\, \overset{\triangledown}{val} / i \,]$$

A third operation on Rep_{Env}, called "label", will be added later.

Value Representation.

Rep_{Val}'s record format is [*tag lft rgt env*], and includes boolean values [BIT *b*], numerals [NUM *n*], error messages [ERR *m*], and primitive operations [OPR *o*]. Other value-objects which use the *rgt* and *env* fields will be added later.

5.3.3. Definition of IM.

We define the concrete interpreter IM by cases, according to expression type. In presenting the definition we first write down IM's intended abstract value, and then look for an expression in reduced terms that has that value. We may have to introduce new objects with special properties to succeed. Existence of these objects is assumed. The presentation can later be viewed as a proof of IM's partial correctness, depending on the existence of the postulated objects.

Numerals. We intend

$$^{\triangledown}[\mathrm{IM}([\mathrm{NUM}\ n],\ env)] = \mathbb{L}\ [\![\ n\ ]\!]\ \overset{\triangledown}{env} = n.$$

Assuming that $^{\triangledown}[\mathrm{NUM}\ n] = n$, define

$$\mathrm{IM}([\mathrm{NUM}\ n],\ env) = [\mathrm{NUM}\ n].$$

Identifiers. We intend

$$^{\triangledown}[\mathrm{IM}([\mathrm{IDE}\ i],\ env)] = \mathbb{L}\ [\![\ i\ ]\!]\overset{\triangledown}{env} = \overset{\triangledown}{env}\ (i).$$

Since we have already assumed that $^{\triangledown}[\mathrm{find}(i,\ env)] = \overset{\triangledown}{env}(i)$, we should define

$$\mathrm{IM}([\mathrm{IDE}\ i],\ env) = \mathrm{find}(i,\ env).$$

λ-expressions. We intend

$$^{\triangledown}[\mathrm{IM}([\mathrm{LAM}\ i\ exp],\ env)] = \mathbb{L}\ [\![\ \lambda\ i\ .\ exp\ ]\!]\ \overset{\triangledown}{env} = \lambda\ v\ .\ \mathbb{L}\ [\![\ \overset{\triangledown}{exp}\ ]\!]\ (\overset{\triangledown}{env}\ [v/i])$$

and need something in Rep_{Val} to stand for the right-hand object. Add *function closures* to Rep_{Val} with record format $[\mathrm{FTN}\ i\ exp\ env]$. If we can ensure that

$$^{\triangledown}[\mathrm{FTN}\ i\ exp\ env] = \lambda\ v\ .\ \mathbb{L}[\overset{\triangledown}{exp}\ ]\ (\overset{\triangledown}{env}\ [v/i])$$

then we can define

$$\mathrm{IM}([\mathrm{LAM}\ i\ exp],\ env) = \text{make-FTN}(i\ exp\ env).$$

A function closure adequately represents its abstraction if it produces the right answer whenever it is applied. An agent is needed to do application. Define

$$\mathrm{APPLY}([\mathrm{FTN}\ i\ exp\ env],\ val) = \mathrm{IM}(exp,\ \mathrm{extend}(env,\ val,\ i)).$$

Then by earlier assumptions,

$$^\triangledown[\text{APPLY } ([\text{FTN } i \; exp \; env], \; val\,)]$$

$$= \;^\triangledown[\text{IM}(exp\,, \text{extend}(env\,, \; val\,, \; i))] \qquad \Delta \; APPLY$$

$$= \; \mathbb{L} [\![\; \overset{\triangledown}{exp} \;]\!] (^\triangledown[\text{extend}(env\,, \; val\,, \; i)]) \qquad Induction \; Hypothesis$$

$$= \; \mathbb{L} [\![\; \overset{\triangledown}{exp} \;]\!] \, (\overset{\triangledown}{env} \, [\overset{\triangledown}{val} \, / i]) \qquad Assumption \; about \; extend$$

$$= (\; \lambda \, v \, . \; \mathbb{L} [\![\; \overset{\triangledown}{exp} \;]\!] \overset{\triangledown}{env} \, [v \, / i]) \overset{\triangledown}{val} \qquad substitution$$

as desired.

Label-expressions. To avoid dealing directly with the *fix* operation we shall hide it in the environment specification. Let us modify the original definition of $\mathbb{L}$.

PROPOSITION 5.3-1. For $\alpha : Env \to Val$,

$$fix \, (\, \lambda \epsilon . \alpha \rho [\, \epsilon \, / \, i \,] \,) = \alpha \; (fix \, (\, \lambda \rho' . \rho [\, \alpha \rho' \, / \, i \,] \,)).$$

PROOF: (Appendix C).

□

COROLLARY 5.3-2 If $\mathbb{L}$'s definition is revised to read

$$\mathbb{L} [\![\; i \Leftarrow e \;]\!] \rho = \mathbb{L} [\![\; e \;]\!] \; (fix \, \lambda \, \rho' \, . \, \rho [\, \mathbb{L} [\![\, e \,]\!] \rho' \, / \, i])$$

the resulting valuation is unchanged.

PROOF: by structural induction on the language L. The valuation is unchanged on base expressions, that is, numerals and identifiers. On composite expressions we may assume by induction that subexpressions have the same valuation. The only questionable case is for "$i \Leftarrow e$", which holds by Proposition 3.5-1 with $\alpha = \mathbb{L} [\![\, e \,]\!]$.

□

Reading $\mathbb{L}$'s new definition literally (if somewhat purposefully), to evaluate "$i \Leftarrow e$" we must arrange to create an environment ρ' that binds i to "the evaluation of e in ρ' ". Hence, a representation is needed for *an evaluation*. Define an *expression closure* to be a value of the form [SPN $exp \; env$]. If we intend $^\triangledown$[SPN $exp \; env$] to equal $\mathbb{L} [\![\; \overset{\triangledown}{exp} \;]\!] \overset{\triangledown}{env}$, then an agent like APPLY is needed to ensure this relationship. We are in the process of defining that agent right now; it is IM. We also need an operation

$$label : (Ide \times Rep_{Exp} \times Rep_{Env}) \to Rep_{Env}$$

that satisfies[2]

$$^{\triangledown}[\text{label}(i, exp, env)] = fix\,\lambda\,\rho' \cdot \overset{\triangledown}{env}[IL[\![\overset{\triangledown}{exp}]\!]\rho'/i].$$

The label-operation is correct if the environment it creates binds the right value to every identifier. Suppose that

$$\text{find}(i, \text{label}(i, exp, env))$$
$$= [\text{SPN}\ exp\ \text{label}(i, exp, env)].$$

Modify IM to test for expression closures whenever it looks to the environment.

$$IM([IDE\ i\,], env) = COERCE(\text{find}(i, env)).$$

$$COERCE([\text{SPN}\ exp\ env']) = IM(exp, env').$$

Then

$$
\begin{array}{ll}
^{\triangledown}[IM(i, \text{label}(i, exp, env))] & \\
= {}^{\triangledown}[IM([\text{SPN}\ exp\ \text{label}(i, exp, env)], \text{label}(i, exp, env))] & \Delta\ \text{label}, IM \\
= {}^{\triangledown}[IM(exp, \text{label}(i, exp, env))] & \Delta\ COERCE \\
= IL[\![\overset{\triangledown}{exp}]\!]^{\triangledown}[\text{label}(i, exp, env)] & I.H. \\
= IL[\![\overset{\triangledown}{exp}]\!](fix\,\lambda\,\rho' \cdot \overset{\triangledown}{env}[(IL[\![\overset{\triangledown}{exp}]\!]\rho'/i]). & \text{intention of label}
\end{array}
$$

Therefore, define

$$IM([LBL\ i\ exp\,], env) = IM(exp, \text{label}(i, exp, env)).$$

Applications. We intend

$$^{\triangledown}[IM([APL\ exp_1\ exp_2], env)] = IL[\![\overset{\triangledown}{exp_1} : \overset{\triangledown}{exp_2}]\!]\overset{\triangledown}{env}$$
$$= apply\,(IL[\![\overset{\triangledown}{exp_1}]\!]\overset{\triangledown}{env})\,(IL[\![\overset{\triangledown}{exp_2}]\!]\overset{\triangledown}{env})$$

We shall implement *apply* by completing the specification of *APPLY* begun earlier. In case that the exp_1 evaluates to a function closure, we already specified how it should be applied when we looked at lambda expressions. Presumably,

---

[2]The implementation in Appendix B defines

label(*ide, exp, env*) $\Leftarrow$ **rec** x **where** x = extend(*env*, make-SPN(*exp*, x), *ide*)).

Thus, we once again build a self-referential representation for the recursive specification. For a recent discussion of this issue, see (Wand, 1983).

the machine has the underlying capability to apply operators. That is, assume there is a mechanism, "apply", such that

$$^{\triangledown}[\text{apply}([\text{OPR } o], val\,)] = o\,(v\overset{\triangledown}{al}\,).$$

Any other value produces an error when applied. The following definition of APPLY accounts for all the cases:

$$\text{APPLY}(ftn,\ arg\,) \;\Longleftarrow$$
$$\text{opr?}(ftn\,) \rightarrow \text{apply}(ftn, arg\,),$$
$$\text{ftn?}(ftn\,) \rightarrow \textbf{let}\ [tag\ ide\ exp\ env] = ftn$$
$$\textbf{in}\ \text{IM}(exp,\ \text{extend}(env,\ ide,\ val\,)),$$
$$\text{make-ERR}(\text{"}invalid\ function\text{"}).$$

The DAISY-like declaration "**let** $[tag\ ide\ exp\ env] = ftn$" simply states $ftn's$ record structure in the case that it is a function closure. Subsequent occurrences of the field names could be replaced by the corresponding field extraction operations.

Conditionals. We intend

$$^{\triangledown}[\text{IM}([\text{CND } exp_1\ [\text{TST } exp_2\ exp_3]\,]] = \mathbb{L}\ [\![\ e\overset{\triangledown}{xp}_1 \rightarrow e\overset{\triangledown}{xp}_2,\ e\overset{\triangledown}{xp}_3\ ]\!]\ e\overset{\triangledown}{nv}$$

On the right we get

$$test\,(\mathbb{L}\ [\![\ e\overset{\triangledown}{xp}_1\ ]\!]\ e\overset{\triangledown}{nv}\,) \rightarrow (\mathbb{L}\ [\![\ e\overset{\triangledown}{xp}_2\ ]\!]\,e\overset{\triangledown}{nv}\,),\ (\mathbb{L}\ [\![\ e\overset{\triangledown}{xp}_3\ ]\!]e\overset{\triangledown}{nv}\,).$$

Assume there is a primitive operation, $test : Rep_{Val} \rightarrow Bool$, that satisfies $test(val\,) = test\,(v\overset{\triangledown}{al}\,)$, and define

$$\text{IM}([\text{CND } exp_1\ [\text{TST } exp_2\ exp_3]\,],\ env\,) =$$
$$test(\text{IM}(exp_1,\ env\,)) \rightarrow \text{IM}(exp_2,\ env\,),\ \text{IM}(exp_3,\ env\,).$$

This completes the construction of a concrete specification for the L-interpreter. Two new types have been added to Rep_{Val}: function closures and expression closures. Thus, the possible value records are:

operator – $[\text{OPR } o\,]$	*error message* – $[\text{ERR } m\,]$
numeral – $[\text{NUM } n\,]$	*function closure* – $[\text{FTN } i\ exp\ env\,]$
boolean – $[\text{BIT } b\,]$	*expression closure* – $[\text{SPN } exp\ env\,]$

Figure 5.4 gives the specification of IM from the discussion above. We have postulated an underlying type that includes the representations, representation

builders, field extraction primitives, and operations find, extend, label, apply, and test.

$\text{IM}(\ exp,\ env\ ) \Longleftarrow$

    let $[tag\ lft\ rgt\ ] = exp$
    in
   num?(exp) $\rightarrow$ $exp,$
    ide?(exp) $\rightarrow$ COERCE(find($lft,\ env$)),
   lam?(exp) $\rightarrow$ make-FTN($lft,\ rgt,\ env$),
    lbl?(exp) $\rightarrow$ $\text{IM}(\ rgt,$ label($lft,\ rgt,\ env$)),
   apl?(exp) $\rightarrow$ APPLY($\text{IM}(\ lft,\ env$), $\text{IM}(\ rgt,\ env$)),

   cnd?(exp) $\rightarrow$ **let** $[tag'\ lft'\ rgt'\ ] = rgt$
                  in test($\text{IM}(\ lft,\ env$)) $\rightarrow$ $\text{IM}(\ lft',\ env$), $\text{IM}(\ rgt',\ env$).

COERCE(val) $\Longleftarrow$

   opr?(val) $\rightarrow$ $val,$
  num?(val) $\rightarrow$ $val,$
    err?(val) $\rightarrow$ $val,$
    ftn?(val) $\rightarrow$ $val,$

   spn?(val) $\rightarrow$ **let** $[tag\ exp'\ env'\ ] = val$
                 in $\text{IM}(\ exp,\ env$).

APPLY($ftn,\ arg$) $\Longleftarrow$

   opr?(ftn) $\rightarrow$ apply($ftn,\ arg$),

  ftn?(ftn) $\rightarrow$ **let** $[tag\ ide\ exp\ env\ ] = ftn$
                in $\text{IM}(\ exp,$ extend($ide,\ arg,\ env$)),
         make-ERR(*"invalid function"*).

Figure 5.4. Non-linear Specification for an L-interpreter

5.3.4. Stacking Version of IM.

Using the Wand-Friedman construction discussed in Section 2.4.5, IM is now transformed to an iterative specification with a control stack. The result is shown in Figure 5.5. Since the construction forces us to choose an evaluation order for recursive calls, we end up at a weaker interpreter than the formal definition demands. In this case an "applicative order" interpreter is derived. For example, the L-expression $[\![\, 5 : (x \Leftarrow x)\,]\!]$ should produce an error message according to the definition of L, and does so under the IM of Figure 5.4. However, its interpretation diverges under the IM of Figure 5.5 (See the last experiment in Appendix B).

In this case, an appropriate version of the control stack is one on which environments and *actions* can be pushed. Actions are represented by records of the form [*tag lft rgt*]. The possible actions are

$$[\, \text{HLT} \,] \text{ — } halt \text{ the machine}$$
$$[\, \text{ARG } exp \,] \text{ — evaluate an application's } arg\text{ument}$$
$$[\, \text{ACT } val \,] \text{ — to apply a function}$$
$$[\, \text{TST } exp_1 \; exp_2 \,] \text{ — } test \text{ a conditional's predicate.}$$

We have allowed the right subfield of a CND-type expression, always something of the form [TST exp_1 exp_2], to serve literally as an action, so the trivial translator R is something of a compiler after all.

```
IM( exp, stk, env ) ⇐
  let [tag lft rgt ] = exp
  in num?( exp ) → RETURN( exp, stk ),
      ide?( exp ) → COERCE(find( lft, env ), stk ),
      lam?( exp ) → RETURN(make-FTN( lft, rgt, env ), stk ),
      lbl?( exp ) → IM( rgt, stk, label( lft, rgt, env )),
      apl?( exp ) → IM( lft, push(make-ARG( rgt ), env, stk ), env ),
      cnd?( exp ) → IM( lft,, push( rgt, env, stk ), env ).


COERCE( val, stk ) ⇐
  let [tag exp env ] = val
  in  opr?( val ) → RETURN( val, stk ),
      num?( val ) → RETURN( val, stk ),
      err?( val ) → RETURN( val, stk ),
      ftn?( val ) → RETURN( val, stk ),
      spn?( val ) → IM( exp, stk, env ).


RETURN( val, stk ) ⇐
  let [nxt env ] = top( stk )
      [tag lft rgt ] = nxt
      stk' = pop( stk )
  in  hlt?( nxt ) → val ,
      tst?( nxt ) → [(test( val ) → IM( lft, stk', env ), IM( rgt, stk', env )],
      arg?( nxt ) → IM( lft, push(make-ACT( lft ), , stk' ), env ),
      act?( nxt ) → APPLY( lft, val, stk' ).


APPLY( ftn, arg, stk ) ⇐
  let [ tag ide exp env ] = ftn
  in  opr?( ftn ) → RETURN(apply( ftn, arg), stk ),
      ftn?( ftn ) → IM( exp, stk, extend( env, ide, arg )),
              RETURN(make-ERR( "invalid function" ), stk ).
```

Figure 5.5. Stacking Version of the L-interpreter.

5.3.5. Simple Loop for the L-interpreter. We now use the construction of Section 2.4.3 to compile the IM of Figure 5.5 into the simple loop shown in Figure 5.6. To prepare for the transformation, all of the serious functions must be defined over the same state descriptor. The various argument names are combined to a single formal parameter list, and the defining equations are altered appropriately. The functions modify only those parameters they originally depended on, and pass the arbitrary value ϕ, in the other positions.

A control token c is added to encode which of IM (**E**, for "EVAL"), COERCE (**C**), APPLY (**A**), or RETURN (**R**) is in control. In the case that c equals **R** and the action is a test, the selection of an alternative expression is distributed through the recursive call to IM. That is, we have changed the clause

$$\text{test}(\text{ val }) \rightarrow \text{IM}(\textbf{E}, \phi, \phi, \phi, \text{lft}', \text{stk}', \text{old}), \text{IM}(\textbf{E}, \phi, \phi, \phi, \text{rgt}', \text{stk}', \text{old})$$

to

$$\text{IM}(\textbf{E}, \phi, \phi, \phi, [\text{test}(\text{ val }) \rightarrow \text{lft}', \text{rgt}'], \text{stk}', \text{old}).$$

We are safe in making this local transformation since the system is linear and the conditional involves only total operations.

5.3.6. Some Refinements in the Loop Version. A little design refinement is irresistible. Note the following about the machine of Figure 5.6.

1. Unless an identifier is bound to an expression closure, its evaluation results in simply moving its binding to position *val* and returning.

2. There are only three cases when a type predicate is used in two states. The predicate *num?* is used at EVAL and COERCE The predicates *opr?* and *ftn?* are used at COERCE and APPLY

3. When control is at RETURN, the argument *exp* is not used.

4. The arguments *ftn* and *val* are unused except when control passes to APPLY, and in APPLY the arguments *exp* and *val* are unused.

With these points in mind, let us now propose that expressions, values, and actions be "superimposible", like variant records. That is, suppose they are represented in such a way as to be accessed by the same field extraction primitives. This allows us to do some register optimization. (The trick of allowing **TST**-expressions to serve as actions foreshadows this refinement.) If the tags are kept distinct, we can make several local transformations on IM that reduce the

```
IM( ctl, ftn, arg, val, exp, stk, env ) ⟸
  let
     [tage ide f-text f-env] = ftn
     [tag v-text v-env] = val
     [tag lft rgt] = exp
     [nxt old] = top( stk)
     [tag lft´ rgt´] = nxt
     stk´ = pop( stk )
  in
    ( ctl = E) →
   num?( exp ) → IM( R, φ, φ, exp, φ, stk, env ),
    ide?( exp ) → IM( C, φ, φ, find( exp, env ), φ, stk, env ),
   lam?( exp ) → IM( R, φ, φ, make-FTN( lft, rgt, env ) , φ, stk, env ),
    lbl?( exp ) → IM( E, φ, φ, φ, rgt, stk, label( lft, rgt, env )),
    apl?( exp ) → IM( E, φ, φ, φ, lft, push(make-ARG( rgt ), env, stk ), env ),
    cnd?( exp ) → IM( E, φ, φ, φ, lft, push( rgt, env, stk ), stk ),
    (ctl = C) →
    err?( val ) → IM( R, φ, φ, val, φ, stk, env ),
   num?( val ) → IM( R, φ, φ, val, φ, stk, env ),
    opr?( val ) → IM( R, φ, φ, val, φ, stk, env ),
    ftn?( val ) → IM( R, φ, φ, val, φ, stk, env ),
    spn?( val ) → IM( E, φ, φ, φ, v-text, stk, v-env ),
    (ctl = R) →
    hlt?( nxt ) → val ,
    tst?( nxt ) → IM( E, φ, φ, φ, [test(val ) → lft´, rgt´ ], stk´, old ),
    arg?( nxt ) → IM( E, φ, φ, φ, lft´, push(make-ACT( val ), φ, stk´ ), old ),
    act?( nxt ) → IM( A, val, lft´, φ, φ, stk´, old ),
    (ctl = A) →
    opr?( ftn ) → IM( R, φ, φ, apply( ftn, arg ), φ, stk, env ),
    ftn?( ftn ) → IM( E, φ, φ, φ, rgt, stk, extend( env´, lft, val )),
              IM( R, φ, φ, make-ERR("invalid function" ), φ, stk, env ).
```

Figure 5.6. Simple Loop for the L-interpreter.

size of its specification. The result is an equivalent version of IM shown in Figure 5.7.

1. Change IM at EVAL in the case that *exp* is an identifier. Place the term "find(*exp, env*)" back in *exp*. Alter COERCE to test *exp* rather than *val*. Since the only overlap is in the case of numerals, which are handled the same way by COERCE and EVAL...

2. ...combine COERCE and EVAL into a single case.

3. Alter every branch to RETURN to place the top action on the control stack in *exp*. We are simply "spreading" the stack into an available vacant register. If none of the action tags equals any of the expression or value tags, we may also combine the states RETURN and EVAL/COERCE.

4. Use *val* and *exp* to hold the argument and function when going to APPLY.

5.3.7. Realization of IM.
We now have IM expressed as a simple loop and can transcribe it into a circuit description according to Theorem 3.3-5. Components are enclosed in braces to make it easier to discern them from signal identifiers. The entire conditional structure is distributed across the state descriptor, making IM the outermost symbol. Figure 5.8 defines a packaged combination, MUX_M, that implements the required conditional. Figure 5.9 shows the resulting circuit equation.

5.3.8. Refined Realization of IM.
The final transformation, shown in Figure 5.10, factors out complex-typed signals by replacing signals *STK* and *ENV* with abstract components STACK and ENVIRONMENT, defined in Figure 5.8. Both are specialized to serve this circuit. STACK takes instructions **PSH**, **POP**, and **NOP**, and saves actions and environments. ENVIRONMENT takes instructions **SET** to change the environment in effect, **HLD** to keep the current environment, **LAB** to produce a labeled environment, and **EXT** to extend the current environment. It continually *finds* a binding for one of its input signals.

The defining equation for the signal C has been simplified to eliminate one MUX_M component. The circuit goes into APPLY exactly when the expression register holds an action of type ACT. The resulting realization is the last of the derivation.

IM(*ctl, val, exp, stk, env*) $\Longleftarrow$

  let

    [*tag lft rgt env´*] =

    [*nxt old*] = *top(stk*)

    stk´ = pop(*stk*)

  in

    (*ctl* = **E**) →

      hlt?(*exp*) → *val* ,

      num?(*exp*) → IM(**E**, *exp*, *nxt*, *stk´*, *old*),

      opr?(*exp*) → IM(**E**, *exp*, *nxt*, *stk´*, *old*),

      ide?(*exp*) → IM(**E**, ϕ, find(*exp*, *env*), *stk*, *env*),

      lam?(*exp*) → IM(**E**, make-FTN(*lft*, *rgt*, *env*), *nxt*, *stk´*, *env*),

      lbl?(*exp*) → IM(**E**, ϕ, *rgt*, *stk*, label(*lft*, *rgt*, *env*)),

      apl?(*exp*) → IM(**E**, ϕ, *lft*, push(make-ARG(*rgt*), *env*, *stk*), *env*),

      cnd?(*exp*) → IM(**E**, ϕ, *lft*, push(*rgt*, *env*, *stk*), *env*),

      ftn?(*exp*) → IM(**E**, *exp*, *nxt*, *stk´*, *old*),

      spn?(*exp*) → IM(**E**, ϕ, *lft*, *stk*, *rgt*),

      tst?(*exp*) → IM(**E**, ϕ, [test(*val*) → *lft*, *rgt*], *stk´*, *old*),

      arg?(*exp*) → IM(**E**, ϕ, *lft*, push(make-ACT(*val*), ϕ, *stk*), *old*),

      act?(*exp*) → IM(**A**, *val*, *lft*, *stk´*, *old*),

      err?(*exp*) → IM(**E**, *exp*, *nxt*, *stk´*, *old*),

    (*ctl* = **A**) →

      opr?(*exp*) → IM(**E**, apply(*exp*, *val*), *nxt*, *stk´*, *old*),

      ftn?(*exp*) → IM(**E**, ϕ, *rgt*, *stk´*, extend(*env´*, *lft*, *val*)),

         IM(**E**, make-ERR("*invalid function*"), *nxt*, *stk´*, *old*).

Figure 5.7. Refined Loop for the L-interpreter.

$$\mathbf{mux_M}(c,\ exp, e\text{-}nm,\ e\text{-}opr,\ e\text{-}ide,\ e\text{-}lam,\ e\text{-}lbl,\ \ e\text{-}apl,\ \ e\text{-}cnd,\ e\text{-}ftn,$$
$$e\text{-}spn,\ \ e\text{-}tst,\ \ e\text{-}arg,\ \ e\text{-}act,\ \ e\text{-}err,\ \ a\text{-}opr,\ \ a\text{-}ftn,\ \ a\text{-}err\ ) \Longleftarrow$$

$(\ c = \mathrm{E}\ ) \rightarrow$

$[\ \text{num?}(\ exp\ ) \rightarrow e\text{-}nm\ ,\text{opr?}(\ exp\ ) \rightarrow e\text{-}opr\ ,\text{ide?}(\ exp\ ) \rightarrow e\text{-}ide\ ,\text{lam?}(\ exp\ ) \rightarrow e\text{-}lam\ ,$

$\quad \text{lbl?}(\ exp\ ) \rightarrow e\text{-}lbl\ ,\ \text{apl?}(\ exp\ ) \rightarrow e\text{-}apl\ ,\text{cnd?}(\ exp\ ) \rightarrow e\text{-}cnd\ ,\text{ftn?}(\ exp\ ) \rightarrow e\text{-}ftn\ ,$

$\quad \text{spn?}(\ exp\ ) \rightarrow e\text{-}spn\ ,\ \text{tst?}(\ exp\ ) \rightarrow e\text{-}tst\ ,\ \text{arg?}(\ exp\ ) \rightarrow e\text{-}arg\ ,\ \text{act?}(\ exp\ ) \rightarrow e\text{-}act\ ,$

$\quad \text{err?}(\ exp\ ) \rightarrow e\text{-}err\ ],$

$(\ c = \mathrm{A}\ ) \rightarrow \quad [\ \text{opr?}(\ exp\ ) \rightarrow a\text{-}opr\ ,\ \text{ftn?}(\ exp\ ) \rightarrow a\text{-}ftn,\ a\text{-}err\ ].$

ENVIRONMENT$(e^0,\ \mathrm{INST},\ \mathrm{X},\ \mathrm{Y},\ \mathrm{Z}) \Longleftarrow \mathbf{rec}\ (\ [\text{find}](\mathrm{X},\ \mathrm{ENV}),\ \mathrm{ENV})$

where

$\quad \mathrm{ENV} = e^0\ !\ [\text{mux}_\mathrm{E}](\mathrm{INST},\ \mathrm{ENV},\ \mathrm{X},\ [\text{label}](\mathrm{X},\ \mathrm{Y},\ \mathrm{ENV}),\ [\text{extend}](\mathrm{X}\ \mathrm{Y}\ \mathrm{Z}))$

$\mathbf{mux_E}(inst,\ u,\ v,\ w,\ x\ ) \Longleftarrow$

$\quad (inst = \mathbf{HLD}) \rightarrow u\ ,$

$\quad (inst = \mathbf{SET}) \rightarrow v\ ,$

$\quad (inst = \mathbf{LAB}) \rightarrow w\ ,$

$\quad (inst = \mathbf{EXT}) \rightarrow x\ .$

STACK$(s^0,\ \mathrm{INST},\ \mathrm{ACTN},\ \mathrm{ENV}) \Longleftarrow \mathbf{rec}\ (\mathrm{NXT},\ \mathrm{OLD})$

where

$\quad (\mathrm{NXT},\ \mathrm{OLD}) = \text{transpose}([\text{top}](\mathrm{STK}))$

$\quad \mathrm{STK} = s^0\ !\ [\text{mux}_\mathrm{S}](\mathrm{INST},\ [\text{push}](\mathrm{ACTN},\ \mathrm{ENV},\ \mathrm{STK}),\ [\text{pop}](\mathrm{STK})\ )$

$\mathbf{mux_S}(\ inst,\ u,\ v,\ w\ ) \Longleftarrow$

$\quad (inst = \mathbf{NOP}) \rightarrow u\ ,$

$\quad (inst = \mathbf{POP}) \rightarrow v\ ,$

$\quad (inst = \mathbf{PSH}) \rightarrow w\ .$

Figure 5.8. Higher Level Components for the L-realization.

IM(c^0, v^0, z^0, s^0, e^0) $\Leftarrow$ **rec** *Experiment* **where**

C = c^0 ! [mux$_M$](C, EXP, [E], [E], [E], [E], [E], [E], [E], [E],
    [E], [E], [E], [A], [E], [E], [E], [E])

VAL = v^0 ! [mux$_M$](C, EXP, EXP, EXP, ϕ, CLS, ϕ, ϕ, ϕ, EXP,
    ϕ, ϕ, ϕ, VAL, EXP, ALU, ϕ, ERR)

ALU = [apply](EXP, VAL)

ERR = [make-ERR](EXP, C)

EXP = z^0 ! [mux$_M$](C, EXP, NXT, NXT, FND, NXT, RGT, LFT, LFT, NXT,
    LFT, TST, LFT, LFT, NXT, NXT, RGT, NXT)

LFT = [lft](EXP)

RGT = [rgt](EXP)

TST = [mux]([test](VAL), LFT, RGT)

CLS = [make-FTN](LFT, RGT, ENV)

FND = [find](EXP, ENV)

STK = s^0 ! [mux$_M$](C, EXP, RTN, RTN, STK, RTN, STK, PSH, PSH, RTN,
    STK, STK, PSH, STK, RTN, RTN, STK, RTN)

PSH = [push](ACTN, ENV. STK)

RTN = [pop](STK)

(NXT OLD) = [top](STK)

ACTN = [mux$_M$](C, EXP, ϕ, ϕ, ϕ, ϕ, ϕ, ARG, RGT, ϕ,
    ϕ, ϕ, ACT, ϕ, ϕ, ϕ, ϕ, ϕ)

ARG = [make-ARG](RGT)

ACT = [make-ACT](VAL)

ENV = e^0 ! [mux$_M$](C, EXP, OLD, OLD, ENV, OLD, LBL, ENV, ENV, OLD,
    RGT, ENV, ENV, ENV, OLD, OLD, EXT, OLD)

LBL = [label](LFT, RGT, ENV)

EXT = [extend](LFT, VAL, SAV)

SAV = [env](EXP)

Figure 5.9. Realization of the L-interpreter.

$$\text{IM}(\ c^0,\ v^0,\ x^0,\ s^0,\ e^0\ ) \Longleftarrow \textbf{rec } \textit{Experiment} \textbf{ where}$$

$$C = c^0\ !\ [\text{mux}](\ \text{ACT?}(\text{EXP}),\ [\text{E}],\ [\text{A}])$$

$$\text{VAL} = v^0\ !\ [\text{mux}_M](C,\ \text{EXP},\ \text{EXP},\ \text{EXP},\quad \phi,\ \text{CLS},\quad \phi,\quad \phi,\quad \phi,\ \text{EXP},$$
$$\phi,\quad \phi,\quad \phi,\ \text{VAL},\ \text{EXP},\ \text{ALU},\quad \phi,\ \text{ERR})$$

$$\text{TST} = \quad [\text{mux}]([\text{test}](\text{VAL}),\ \text{LFT},\ \text{RGT})$$

$$\text{ALU} = \quad [\text{apply}](\text{LFT},\ \text{VAL})$$

$$\text{EXP} = x^0\ !\ [\text{mux}_M](C,\ \text{EXP},\ \text{NXT},\ \text{NXT},\ \text{FND},\ \text{NXT},\ \text{RGT},\ \text{LFT},\ \text{LFT},\ \text{NXT},$$
$$\text{LFT},\ \text{TST},\ \text{LFT},\ \text{LFT},\ \text{NXT},\ \text{NXT},\ \text{RGT},\ \text{NXT})$$

$$\text{LFT} = \quad [\text{lft}](\text{EXP})$$

$$\text{RGT} = \quad [\text{rgt}](\text{EXP})$$

$$\text{SAV} = \quad [\text{env}](\text{EXP})$$

$$\text{CLS} = \quad [\text{make-FTN}](\text{LFT},\ \text{RGT},\ \text{ENV})$$

$$\text{ERR} = \quad [\text{make-ERR}](\text{EXP},\ C)$$

$$(\text{NXT OLD}) = \quad \textbf{STACK}(\ s^0,\ \text{S1},\ \text{S2},\ \text{ENV})$$

$$\text{S1} = \quad [\text{mux}_M](C,\ \text{EXP},[\text{POP}],[\text{POP}],[\text{NOP}],[\text{POP}],[\text{NOP}],\ [\text{PSH}],\ [\text{PSH}],[\text{POP}],$$
$$[\text{NOP}],[\text{NOP}],\ [\text{PSH}],[\text{NOP}],[\text{POP}],\ [\text{POP}],[\text{NOP}],[\text{POP}])$$

$$\text{S2} = \quad [\text{mux}_M](C,\ \text{EXP},\quad \phi,\quad \phi,\quad \phi,\quad \phi,\quad \phi,\ \text{RGT},\ \text{ARG},\quad \phi,$$
$$\phi,\quad \phi,\ \text{ACT},\quad \phi,\quad \phi,\quad \phi,\quad \phi,\quad \phi)$$

$$\text{ARG} = \quad [\text{make-ARG}](\text{RGT})$$

$$\text{ACT} = \quad [\text{make-ACT}](\text{VAL})$$

$$(\text{FND ENV}) = \quad \textbf{ENVIRONMENT}(\ e^0,\ \text{E1},\ \text{E2},\ \text{E3},\ \text{SAV})$$

$$\text{E1} = \quad [\text{mux}_M](C,\ \text{EXP},\ [\text{SET}],\ [\text{SET}],\ [\text{HLD}],\ [\text{SET}],\ [\text{LAB}],\ [\text{HLD}],\ [\text{HLD}],\ [\text{SET}],$$
$$[\text{SET}],\ [\text{HLD}],\ [\text{HLD}],\ [\text{HLD}],\ [\text{SET}],\ [\text{SET}],\ [\text{EXT}],\ [\text{SET}])$$

$$\text{E2} = \quad [\text{mux}_M](C,\ \text{EXP},\ \text{OLD},\ \text{OLD},\ \text{LFT},\ \text{OLD},\ \text{LFT},\quad \phi,\quad \phi,\ \text{OLD},$$
$$\text{RGT},\quad \phi,\quad \phi,\quad \phi,\ \text{OLD},\ \text{OLD},\ \text{RGT},\ \text{OLD})$$

$$\text{E3} = \quad [\text{mux}_M](C,\ \text{EXP},\quad \phi,\quad \phi,\quad \phi,\quad \phi,\ \text{RGT},\quad \phi,\quad \phi,\quad \phi,$$
$$\phi,\quad \phi,\quad \phi,\quad \phi,\quad \phi,\quad \phi,\ \text{VAL},\quad \phi)$$

Figure 5.10. Refined L-realization.

5.3.9. Remarks. We have derived a description for a machine that interprets suitably represented expressions in the language L with a call-by-value semantics. In Appendix B, each step in the derivation is expressed in DAISY, and a set of trial expressions are interpreted by the various versions of IM.

The executable versions of IM's specification would eventually serve as an experimental vehicle for continued design refinement. For example, a trace of the circuit shows that it wastes cycles testing for expression closures. (See the last experiment in Appendix B.) We would do better to make that test concurrent with evaluation, so that the presence of an expression closure has the effect of an interrupt. Of course, innumerable other modifications are possible, and we shall not pursue them here.

The programs in Appendix B were used to debug the derivation. Since the transformations were carried out by hand, there were a number of errors. Many were discovered by attempting to execute the erroneous forms.

In transforming the specification to one having linear control, L's semantics have been weakened; there are expressions which converge under the initial specification but do not on the target machine. We could alter L's formal specification to reflect this change in its design. Figure 5.11, giving L's continuation semantics, is the appropriate modification. While we took a separate step to introduce the control stack, the transformation is entirely in the spirit of Section 5.3.2. Had we started with L's continuation semantics rather than its standard semantics, we would have proposed a representation for continuations and introduced the appropriate agents and operations for these objects directly (Wand, 1982a).

<u>Domains</u>

Ide	*(i)*	*identifiers*
Num	*(n)*	*numerals*
Bool	*(b)*	*truth values*
Opn = Val → Bas	*(o)*	*operations*
Err = { "invalid function",...}	*(m)*	*error messages*
Exp	*(e)*	*expressions*
Bas = Num + Bool + Opn + Err	*(v)*	*basic values*
Val = Bas + Ftn + Spn	*(v)*	*expressible values*
Ftn = Val → Spn	*(f)*	*functions*
Spn = K → Val	*(σ)*	*expression closures*
K = Val → Val	*(κ)*	*expression continuations*
Env = Ide → Val	*(ρ)*	*environments*

<u>Valuation</u> – $\mathbb{L} : Exp \rightarrow Env \rightarrow K \rightarrow Val$

$$\mathbb{L} [\![\, n \,]\!] \rho\kappa = \kappa\, n$$
$$\mathbb{L} [\![\, i \,]\!] \rho\kappa = coerce\ (\rho i)\ \kappa$$
$$\mathbb{L} [\![\, \lambda\, i \,.\, e \,]\!] \rho\kappa = \kappa\ (\lambda v\kappa' \,.\ \mathbb{L} [\![\, e \,]\!]\ (\rho[v/i])\ \kappa')$$
$$\mathbb{L} [\![\, i \Leftarrow e \,]\!] \rho\kappa = fix\ (\lambda\epsilon.\ \mathbb{L} [\![\, e \,]\!]\ \rho[\epsilon/i])$$
$$\mathbb{L} [\![\, e_1 : e_2 \,]\!] \rho\kappa = \mathbb{L} [\![\, e_1 \,]\!]\rho\ (\lambda f.\mathbb{L} [\![\, e_2 \,]\!]\rho\ (\lambda v.\ (apply\ f\ v\ \kappa\ )))$$
$$\mathbb{L} [\![\, e_1 \rightarrow e_2,\, e_3 \,]\!] \rho\kappa = \mathbb{L} [\![\, e_1 \,]\!]\rho\ (\lambda v.\ (test\ v) \rightarrow \mathbb{L} [\![\, e_2 \,]\!]\rho\kappa,\ \mathbb{L} [\![\, e_3 \,]\!]\rho\kappa\ )$$

<u>Auxiliaries</u>

$$\rho[v/i] = \lambda x.\ (x = i) \rightarrow v,\ \rho(x).$$
$$coerce = \lambda v\kappa.\ (v\ isSpn) \rightarrow (v\ \kappa),\ (\kappa\ v)$$
$$apply = \lambda fa\kappa.\ (f\ isOpr) \rightarrow \kappa(fv),\ (f\ isFtn) \rightarrow fv\kappa,\ \text{``invalid function''}.$$
$$test = \lambda v.\ (v\ isBool) \rightarrow (v\ asBool),\ \mathit{ff}.$$

Figure 5.11. Continuation Semantics for L.

6. Circuit Refinement

Experiments with realizations in Section 4.4.3 and Chapter 5 have revealed that the derived circuits can be improved. In this chapter we turn to the issue of refining circuit descriptions. Although a specific refinement task is considered below, the method of refinement is consistent with the methods developed earlier. A specialized set of transformation rules is used to attain a goal. Since we are concerned with improving circuits and not deriving them, both source and target descriptions will now be connectivity expressions. The initial specification describes the instantaneous behavior of a combined operation. Our transformations yield digital system realizations that perform the same computation as the specification but do it in a serial fashion. Of course this complicates the timing of the circuit involved.

The task is to modify a large combinatorial system so that it has fewer external connections. This goal is attained by "folding" the system so that components, and hence external leads, are superimposed. Since an individual component cannot simultaneously produce two results, it is necessary to serialize its behavior. Time is traded against space, where the latter is measured in terms of a "pin count".

It will be necessary to keep track of individual values produced by the system in circuit-folding derivations. Since DAISY is fairly useful for this kind of bookkeeping, it is used as the transformation medium. That is, we shall build our algebra of synthesis on DAISY-like notation, rather than the purely functional or purely sequential languages used earlier.

To illustrate the problem, let us consider the following configuration of components.

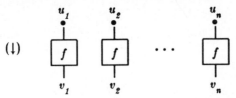

The operation f is applied in parallel and independently to the individual values $u_1, ..., u_n$, producing results $v_1, ..., v_n$. Suppose that we want to implement a design that has only one input and one output. The obvious modification is to serialize the u's and use a single f-component:

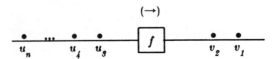

The schematic notation above is informal. The series of tokens along a single wire simply illustrates that the f-component is acting on each of the u's. One should not read too much into this picture; for example, it does not imply that wires necessarily store values. The price for reducing the external connectivity of this system is that the surrounding circuit must somehow be modified to support the serialization. This is a simple serialization problem; the u's can be presented in any order; the v's are produced in the same relative order. Now consider a system that has internal connectivity:

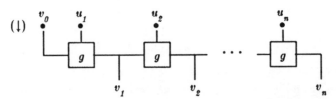

The g's can be superimposed as before, but in this case the attempt at folding introduces feedback. A register[1] is needed to synchronize the system internally.

---

[1] This is a good time to recall that the word "register" was adopted for its brevity and to note that what is really meant here is *storage element*. How storage is implemented depends on the fabrication medium.

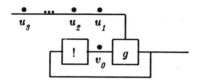

The surrounding circuit must present the *u's* in an order that exploits the feedback in the circuit. There is only one suitable ordering.

In seeking a method of synthesizing *local* refinements like the ones above, we should, if possible, account for the changing performance relationship between the subcircuit and the system that surrounds it. We develop a transformation system below in which, at some point during a derivation, we elect to identify some fragment of the evolving description as a signal. Once the identification is made, the content of the signal becomes superfluous in the description. However, this residual information is retained as a specification of external behavior and can be used as a basis for coordinating the target system with its surroundings.

The order that values are serialized drives the transformation process. While the examples in this chapter are small enough that appropriate orderings can be deduced by inspection, it will be apparent that without guiding heuristics synthesis would be hopelessly explosive. Gannon (1982) discusses a method to analyze regularly connected data-flow systems to find appropriate orderings. His model assumes connective storage: process coordination is achieved by storage along connecting paths. Both examples above accept tokens on such a path, and deliver values to another. Note however, that some of the connective storage is already implemented if the *g*-circuit's output is taken from its internal register:

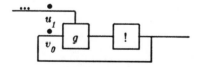

Having introduced storage we now need a clock. In transforming a combinatorial subsystem to a synchronous one, it becomes necessary to determine how the target's temporal behavior can be coordinated with it surroundings. Cuny and Snyder (1982) present a model in which autonomous processing elements are specified according to their external communicative behavior. and they address the problem of finding viable computation rates by which the processors can be interleaved to perform synchronously. A process is described by a regular expression over sets of transactions. We shall call such an expression a *schedule*. In their notation, our second example above initially had schedule

$$[\{R_{v_0}, R_{u_1},...,R_{u_n}, W_{v_1},..., W_{v_n} \}]^*$$

(The subscripts identify external connections. R means "read"; W means "write". In the work cited, connections are identified by the names of the surrounding processing elements. These names are not known here, so port identifiers are used instead.)

The initial system is combinatorial; it does all its transactions at once. The refined version is sequential, either

$$[\{R_{v_0} R_U W_V\} [\{R_U W_V\}]^{n-1}]^*$$

or

$$[\{R_{v_0} R_U \} [\{R_U W_V\}]^{n-1} \{W_V\}]^*$$

depending on whether or not the internal register is used to buffer output. The residual byproduct of synthesis mentioned above will be displayed in a form from which such expressions could be extracted.

The synthesis method developed in the rest of this chapter does not compete with methods such as Gannon's or Cuny's and Snyder's, rather it serves as a bridge between them. On large problems analysis is needed both to guide the construction of solution circuits and to deal with the increased temporal complexity of the target system. The method offers a way to maintain correctness while constructing realizations that achieve the goal of serialization. Section 6.1 develops a set of basic transformations on DAISY expressions. These are generalized in Section 6.2 to a rewriting system that we shall use to attack serialization problems. Section 6.3 presents three examples of "scheduling derivations" on increasingly complex configurations of components.

6.1. Transformation Axioms

A combinatorial system will be specified in DAISY by a system of value-defining equations (Sec. 4.2). The left-hand sides of these equations are *formal expressions:* identifier structures delimited by square braces '[' and ']'. The right-hand sides are *actual expressions:* value structures delimited by angle brackets '$<$' and '$>$'. To evaluate these systems, an environment is constructed that recursively binds formal structures to values. For the remainder of this section suppose we are dealing with a specification

$$S \Longleftarrow \mathbf{rec}\ e\ \mathbf{where}$$
$$x_1 = a_1$$
$$\cdot$$
$$\cdot$$
$$\cdot$$
$$x_n = a_n$$

where experiment e is an expression over $x_1,..., x_n$. Recall that S's value is

$$I\!\!D[\![\ e\ ]\!]\ fix\ (\ \lambda\rho'\ .\ \rho_0[\ {}^{I\!\!D[\![\ <a_1 ...a_n>\ ]\!]\rho'}\ /\ {}_{[x_1...x_n]}\ ]\ )$$

where ρ_0 is some initial environment (Sec. 4.3). Let $\downarrow\! a$ denote the value of a in ρ'; that is, $\downarrow\! a = I\!\!D[\![\ a\ ]\!]\rho'$. Then by DAISY's version of environment extension,

$$\rho' = \rho_0[\ \downarrow\! a_n\ /\ x_n\ ]\ ...\ [\ \downarrow\! a_1\ /\ x_1\ ].$$

However, ρ_0 can be extended in any order, as long as the x's are distinct, and in fact can be arbitrarily restructured. With this in mind, we propose the following axioms for transformations on S:

AxiomN: *(Vacuous Equations) The equation "* $[\ ] = <>$ *" can be added to S.*

AxiomG: *(Gluing) The equations "$x = a$" and "$y = b$" can be replaced by* "$[\ x\ !\ y\ ] = <a\ !\ b>$".

AxiomE: *(Extraneous Equations) If the identifier y is free in S then*
(i) The equation "$y = b$" can be added to S, for any expression b.
(ii) The equation "$x = a$" can be replaced by "$x = y$" and "$y = a$".

AxiomF: *(Function Factoring) The expression "$<f\!:\!a_1\ ...\ f\!:\!a_m>$" can be replaced by "$<f\ ...\ f\ >\!:\!\Xi\!:\!<a_1\ ...\ a_m>$", where Ξ stands for a transposition operation.*

AxiomS: *(Signal Interpretation) $Nil = <\phi\ *>$.*

These axioms are all valid in the semantics of DAISY. Axiom **N** introduces an equation that has no effect in S because the formal expression contains no identifiers. Axioms **G** and **E** do not change the value of S because the list constructor is not strict; adding unused bindings and indirection through extraneous names simply restructures the environment. Axiom **F** exploits DAISY's application combinator, *d-apply*. The point of the axiom is that operations may be factored out of structures by applying the reduction rule for function-lists in reverse. Axiom **S** foreshadows our intention to interpret some finite sequences as signals: the main goal of synthesis in this chapter is to construct signals by serializing values. Consequently, we shall permit finite sequences to be interpreted as signals with only finitely meaningful prefixes. This is the only axiom whose validity cannot be deduced directly from the definition of DAISY in Chapter 4. In fact, *Nil* is implemented to satisfy $Nil \equiv fix\ \lambda l.\langle \phi\ ,\ l \rangle$, where in DAISY, ϕ (don't-know) is an all-purpose error message.

6.2. General Transformations and their Behavioral Interpretation

We now combine the axioms of the preceding section into a set of specialized transformation rules for the serialization problem. Each definition is followed by a discussion of how the rule makes progress towards a serialized target.

Let the specification S be as before, except that it will now be parameterized by a list of input values:

$$S{:}[u_1 \ldots u_m] \Longleftarrow \mathbf{rec}\ e\ \mathbf{where}$$
$$x_1 = a_1$$
$$\bullet$$
$$\bullet$$
$$\bullet$$
$$x_n = a_n$$

Rule Γ: *(Gluing) Let π be a permutation of $\{1,\ 2,\ ...,\ n\}$. Any subset of equations in $S\ \{x_{\pi(j)} = a_{\pi(j)} \mid 1 \leq j \leq p \leq n\}$ may be rewritten as*
$$[\ x_{\pi(1)}\ x_{\pi(2)}\ \cdots\ x_{\pi(p)}\ ] = <a_{\pi(1)}\ a_{\pi(2)}\ \cdots\ a_{\pi(p)}>.$$

Rule Γ is valid by repeated use of Axiom **G**, as its name suggests. It is used to associate the x's or a's together in a single structure. Often, this structure is later reinterpreted as a signal.

Rule Δ: *(Delay) Equation "$x = a$" may be rewritten as* "$[\,y\,!\,x\,] = <\phi\,!\,a>$", *where y is any formal expression of identifiers that are free in S.*

We obtain the new equation in Δ through Axioms **G** and **E**, by gluing the extraneous equation "$y = \phi$" to "$x = a$". In the derivations below, y will always be a simple identifier and x will always be a linear sequence. So we will be changing equations of the form

$$[x_1\ x_2\ ...] = a$$

to

$$[y\ x_1\ x_2\ ...] = <\phi\,!\,a>.$$

Explicit concatenator symbols indicate that a register has been added to the evolving circuit. This register postpones the x's in time, which is why Δ is called the delay rule. To avoid making up meaningless names, we sometimes write a ϕ for y:

$$[\phi\ x_1\ x_2\ ...] = <\phi\,!\,a>.$$

Rule Λ: *(Lifting) The expression* "$<f{:}a_1\ ...\ f{:}a_n>$" *can be rewritten as* "$<f\,*>{:}\Xi{:}<a_1\ ...\ a_n>$", *where Ξ is the* identity component $<\lambda i.i\,*>$.

The validity of lifting follows from the meaning of application, as discussed in the previous section under Axiom **F**, and from the the interpretation of *Nil* as the everywhere indeterminate signal. The rule differs from Axiom **F** in that here we regard the finite sequence of f's to be a component. By Axiom **F**, "$<f{:}a_1\ ...\ f{:}a_n>$" can be rewritten as "$<f\ ...\ f>{:}\Xi{:}<a_1\ ...\ a_n>$". Extend $<f\ ...\ f>$ to the infinite sequence $<f\,*>$ and by Axiom **S** interpret $<a_1\ ...\ a_n>$ as $<a_1\ ...\ a_n\ \phi\ \phi\ ...>$. The result of application is

$$<f{:}a_1\ ...\ f{:}a_n\ f{:}\phi\ f{:}\phi\ ...>$$

Assuming f completely strict, this becomes

$$< f\text{:}a_1 \dots f\text{:}a_n \; \phi \; \phi \; \dots >$$

which by Axiom **N** we may write as

$$< f\text{:}a_1 \dots f\text{:}a_n >$$

The identity component Ξ does a generalized transposition on inputs of any dimension. This is the same coercion used in the circuit experiments of Section 4.4.3. Rule Λ is used to introduce to the evolving description a single component that serially computes individual values.

Rule **M**: *(Selection) Suppose S contains two equations of the form*

$$[x_1 \; x_2 \dots x_p] = a$$
$$[y_1 \; y_2 \dots y_p] = b$$

Let " $[z_1 \; z_2 \dots z_p]$ " be a formal expression in which z_i is one of x_i or y_i for all i. Then there is a selection component M by which

$$[z_1 \; z_2 \dots z_n] = M : <a \; b>.$$

M is simply a multiplexor with a fixed predicate signal. For example, if we have

$$[x_1 \; x_2] = <a_1 \; a_2>$$
$$[y_1 \; y_2] = <b_1 \; b_2>$$

and we want a signal of the form $[x_1 \; y_2]$, then we may may replace these equations with

$$[x_1 \; y_2] = MUX : <<tt \; ff> \; <a_1 \; a_2><b_1 \; b_2>>$$

We shall denote the fixed predicate signal as a subscript on M, encoded as a string of bits, with $\underline{0}$ interpreted as *tt*.

$$[x_1 \; y_2] = M_{01} : <<a_1 \; a_2><b_1 \; b_2>>$$

Rule Φ: *(Installation) The identifier x is an instance of the value a in S if either a is ϕ or the equation "x = a" can be deduced from S. A formal structure is an instance of an actual structure if its elements are each instances of the corresponding elements of the actual structure. A actual value can be replaced by any of its instances.*

Rule Φ is used to replace values consumed in S with results produced in S. For example, if S contains the equations

$$z = <a\ \phi\ c>$$
$$[x_1\ x_2\ x_3] = <a\ b\ c>$$

By Rule Φ, z's defining equation may be replaced by

$$z = <x_1\ x_2\ x_3>$$

The rule implies that ϕ is truly arbitrary. That is, we must agree that any value may serve where an unknown value is required. Installation is a restricted form of substitution used to introduce feedback. The equations of S do not immediately admit substitution because they are not identities: their left-hand sides are formal structures and their right-hand sides are not. For instance, the defining equation "$x = x$" may bind x to the divergent value[2], just as might the equation "$x = x+1$". We must avoid transformations that would lead to such equations. An naming convention is used to keep track of value instances. The identifier "$\hat{a}$" is by convention an instance of the value a. For example, a name for the parameter u can be introduced by Axiom **E** with an equation of the form "$\hat{u} = u$".

6.3. Scheduling Derivations

We shall give three examples to show how the rules defined in Section 6.2 can be applied to the scheduling problem. The first two come from the discussion in the introduction to this chapter. The third is a somewhat more complicated combination, a portion of a regularly connected network.

---

[2] which is the *minimal* fixed point of the equation. The use of defining equations as identities to reason about LUCID programs leads to the same "glitch" (Ashcroft and Wadge, 1977).

6.3.1. Circuit F. Consider the simple serialization problem for $n = 3$. The combinatorial system is specified:

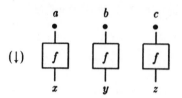

$$S:[a\ b\ c] \Leftarrow \textbf{rec} < x\ y\ z > \textbf{where}$$
$$x = f:<a>$$
$$y = f:<b>$$
$$z = f:<c>$$

From S we may derive

$x = f:<a>$
$y = f:<b>$
$z = f:<c>$

Given

$[x\ y\ z] = <f:<a>\ f:<b>\ f:<c>>$

Γ

$[x\ y\ z] = <f*>:\Xi:<<a><b><c>>$

Λ

$[x\ y\ z] = <f*>:<<a\ b\ c>>$

meaning of Ξ

The final step above is a symbolic transposition of the argument to $<f*>$. We are turning the individual arguments to $<f*>$ into a signal. To emphasize that we are now thinking of S as a digital circuit, let us identify its signals.

$V = <f*>:<U>$
$[x\ y\ z] = V$
$U = <a\ b\ c>$

Identification of signals

The derived circuit applies the combinatorial component $<\!f*\!>$ to its input signal U and produces output signal V. If a, b, and c are presented in order on U then the results x, y, and then z are delivered on V. We shall rephrase this interpretation as a *schedule specification*; the defining equations for U and [x y z] state the external characteristics of the circuit.

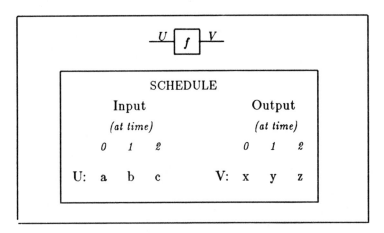

6.3.2. Circuit G. Define S for the simple feedback problem, with $n = 3$.

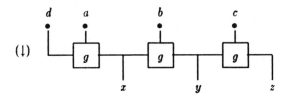

$$S\!:\![d\ a\ b\ c] \Leftarrow \textbf{rec} < x\ y\ z > \textbf{where}$$
$$x = g\!:\!<a\ d>$$
$$y = g\!:\!<b\ x>$$
$$z = g\!:\!<c\ y>$$

We begin by superimposing the *g's* in the only reasonable order.

x = g:\<a d\>	
y = g:\<b x\>	*Given*
z = g:\<c y\>	

$$[x\ y\ z] = <g{:}<a\ d>\ g{:}<b\ x>\ g{:}<c\ y>> \qquad \Gamma$$

$$[x\ y\ z] = <g*>{:}\Xi{:}<<a\ d><b\ x><c\ y>> \qquad \Lambda$$

$$[x\ y\ z] = <g*>{:}<<a\ b\ c>\ <d\ x\ y>> \qquad \Xi$$

Note that x is produced at "time 0" but is consumed at "time 1" by $<g*>$. We shall have to delay this instance of x if the circuit is to use it.

$$[\phi\ x\ y\ z] = \phi\ !\ <g*>{:}<<a\ b\ c>\ <d\ x\ y>> \qquad \Delta$$

Let us invoke our interpretation of *Nil* as the indeterminate signal and write

$$[\phi\ x\ y\ z] = \phi\ !\ <g*>{:}<<a\ b\ c\ \phi>\ <d\ x\ y\ \phi>> \qquad Nil$$

The next goal is to separate the sequence $<d\ x\ y\ \phi>$ into two sequences that segregate internally computed values from externally provided values. Add an extraneous instance of d and do some gluing.

$$\begin{aligned}[\phi\ x\ y\ z] &= \phi\ !\ <g*>{:}<<a\ b\ c\ \phi>\ <d\ x\ y\ \phi>> \\ [\hat{d}\ \phi\ \phi\ \phi] &= <d\ \phi\ \phi\ \phi>\end{aligned} \qquad Axioms\ E,\ G$$

Now by the Selection Rule, the system can be rewritten

$$\begin{aligned}[\hat{d}\ x\ y\ z] &= M_{0111}{:}<<d\ \phi\ \phi\ \phi>\ V> \\ V &= \phi\ !\ <g*>{:}<<a\ b\ c\ \phi>\ <d\ x\ y\ \phi>> \\ [\phi\ x\ y\ z] &= V\end{aligned} \qquad M$$

Since $\hat{d}$ is an instance of d and z is an instance of ϕ, we may rewrite this system as

$$[\hat{d}\ x\ y\ z] = M_{0111}\!:\!<<d\ \phi\ \phi\ \phi>\ V>$$
$$V = \phi\ !\ <g*>\!:\!<<a\ b\ c\ \phi>\ <\hat{d}\ x\ y\ z>>$$
$$[\phi\ x\ y\ z] = V$$

Φ

If we name the rest of the signals we get

$$V = \phi\ !\ g*\!:\!<U\ W>$$
$$W = M_{0111}\!:\!<D\ U>$$

$$[\phi\ x\ y\ z] = V$$
$$U = <a\ b\ c\ \phi>$$
$$D = <d\ \phi\ \phi\ \phi>$$

Identification
of signals

We have derived a description of a circuit of two inputs and one output

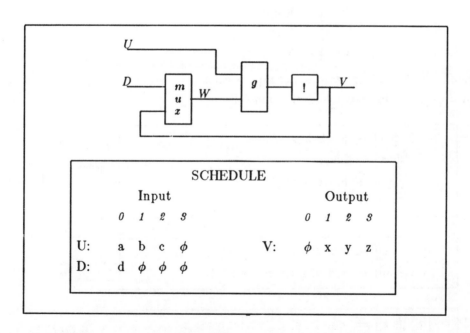

SCHEDULE								
	Input					Output		
	0	*1*	*2*	*3*		*0*	*1*	*2*	*3*
U:	a	b	c	ϕ	V:	ϕ	x	y	z
D:	d	ϕ	ϕ	ϕ					

6.3.3. Circuit H. The final example is a portion of a combinatorial array discussed by Gannon (1982).

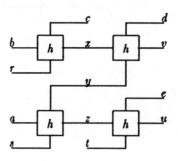

$$S:[a\ b\ c\ d\ e] \Longleftarrow \textbf{rec} <r\ s\ t\ u\ v>$$
where
$$[\ r\ x] = h:<c\ b>$$
$$[\ y\ v] = h:<d\ x>$$
$$[\ s\ z] = h:<y\ a>$$
$$[\ t\ u] = h:<e\ z>$$

As usual, we shall attempt to superimpose certain external leads, namely *a* with *b*, and *d* with *e*. One possible derivation follows:

$$[r\ x] = h:<c\ b>$$
$$[y\ v] = h:<d\ x>$$
$$[s\ z] = h:<y\ a>$$
$$[t\ u] = h:<e\ z>$$

given

$$[[r\ x][s\ z]] = <h:<c\ b>\ h:<y\ a>>$$
$$[[y\ v][t\ u]] = <h:<d\ x>\ h:<e\ z>>$$

Γ, *twice*

$$[[r\ x][s\ z]] = <h*>:<<c\ y>\ <b\ a>>$$
$$[[y\ v][t\ u]] = <h*>:<<d\ e>\ <x\ z>>$$

Λ, *twice*

$$[[r\ s\ \phi][x\ z\ \phi]] = \Xi{:}{<}h*{>}{:}{<}{<}c\ y\ \phi{>}\ {<}b\ a\ \phi{>}{>}$$
$$[[y\ t][v\ u]] = \Xi{:}{<}h*{>}{:}{<}{<}d\ e{>}\ {<}x\ z{>}{>}$$

Ξ, *Nil*

$$[[r\ s\ \phi][x\ z\ \phi]] = \Xi{:}{<}h*{>}{:}{<}{<}c\ y\ \phi{>}\ {<}b\ a\ \phi{>}{>}$$
$$[[y\ t][v\ u]] = \Xi{:}{<}h*{>}{:}{<}{<}d\ e{>}\ {<}x\ z{>}{>}$$
$$[\phi\ \hat{y}\ \hat{t}] = \phi\ !\ {<}y\ t{>}$$
$$[\hat{c}\ \phi\ \phi] = {<}c\ \phi\ \phi{>}$$
$$[\hat{c}\ \hat{y}\ \hat{t}] = M_{011}{:}{<}{<}c\ \phi\ \phi{>}\ {<}\phi\ \hat{y}\ \hat{t}\ {>}{>}$$

(1) Δ, G
(2) Axiom E
M (1) and (2)

$$[[r\ s\ \phi][x\ z\ \phi]] = \Xi{:}{<}h*{>}{:}{<}{<}\hat{c}\ \hat{y}\ \hat{t}{>}\ {<}b\ a\ \phi{>}{>}$$
$$[[y\ t][v\ u]] = \Xi{:}{<}h*{>}{:}{<}{<}d\ e\ \phi{>}\ {<}x\ z\ \phi{>}{>}$$
$$[\phi\ \hat{y}\ \hat{t}] = \phi\ !\ {<}y\ t{>}$$
$$[\hat{c}\ \hat{y}\ \hat{t}] = M_{011}{:}{<}{<}c\ \phi\ \phi{>}\ {<}\phi\ \hat{y}\ \hat{t}\ {>}{>}$$

Φ
Nil

$$[I\ J] = \Xi{:}{<}h*{>}{:}{<}K\ N{>}$$
$$[L\ M] = \Xi{:}{<}h*{>}{:}{<}O\ J{>}$$
$$P = \phi\ !\ L$$
$$K = M_{011}{:}{<}Q\ P{>}$$

$$[r\ s\ \phi] = I$$
$$[x\ z\ \phi] = J$$
$$[y\ t] = L$$
$$[u\ v] = M$$
$$[\phi\ \hat{y}\ \hat{t}] = P$$
$$[\hat{c}\ \hat{y}\ \hat{t}] = K$$
$$N = {<}b\ a\ \phi{>}$$
$$O = {<}d\ e\ \phi{>}$$
$$Q = {<}c\ \phi\ \phi{>}$$

Identification of signals

These equations describe a circuit with external input signals N, O, and Q; external output signals I, M, and P; and internal signals I, K, and L. The schematic and schedule specification are:

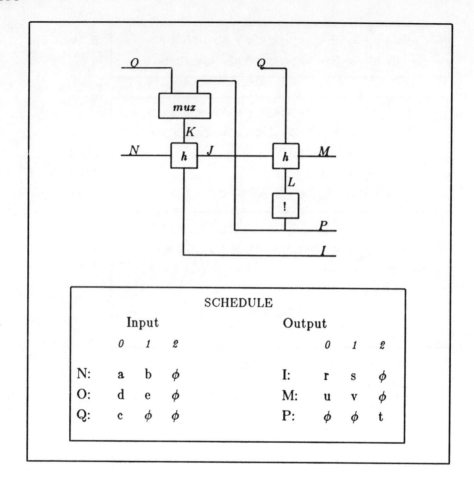

SCHEDULE

	Input					Output		
	0	*1*	*2*			*0*	*1*	*2*
N:	a	b	ϕ		I:	r	s	ϕ
O:	d	e	ϕ		M:	u	v	ϕ
Q:	c	ϕ	ϕ		P:	ϕ	ϕ	t

6.4. Remarks

This chapter demonstrates that the algebraic framework we have developed to *obtain* circuit descriptions is also useful for *refining* them. We introduced a set of axioms and rules that are tailored to a particular problem. The implication is that by similar specialization, a transformation system would evolve to deal with local changes in a design as well as the global generation of one.

The reader may have noticed that in the schedule for Circuit H the stored value t is available earlier if the combinatorial output L is used instead of P. Hence, this circuit can execute its function in two cycles if the register is used solely for internal synchronization. However, even if the delayed occurrence of t is used, the circuit's schedule can be overlapped.

$$
\begin{array}{llllllll}
\text{N:} & a^1 & b^1 & a^2 & b^2 & a^3 & b^3 & \ldots \\
\text{O:} & d^1 & e^1 & d^2 & e^2 & d^3 & e^3 & \ldots \\
\text{Q:} & c^1 & \phi & c^2 & \phi & c^3 & \phi & \ldots \\
\text{I:} & r^1 & s^1 & r^2 & s^2 & r^3 & s^3 & \ldots \\
\text{M:} & v^1 & u^1 & v^2 & u^2 & v^3 & u^3 & \ldots \\
\text{P:} & \phi & \phi & t^1 & \phi & t^2 & \phi & t^3 & \phi & \ldots \\
\end{array}
$$

A single additional cycle is needed to capture the last t. The schedule specifies Circuit H in terms of its external communication. Using the overlapping shown above, its input-output characterization in the Cuny-Snyder notation is

$$\{R_N\,R_O\,R_Q\,W_I\,W_M\}\,[\{R_N\,R_O\,W_I\,W_M\}\,\{R_N\,R_O\,R_Q\,W_I\,W_M\,W_P\}]^*\{W_P\}$$

While each step in these example derivations is a valid transformation on specification text, it is not immediately clear what drives the derivation toward a circuit realization. Our heuristics were to segregate internal from external values, and to introduce delays to align component inputs with component outputs. However, since we are free to introduce delays of any duration, and since some value orderings are inadmissible, a transformation strategy based on those simple heuristics could easily go awry.

In Circuit H one can see that if inputs a and b are serialized, b should precede a because b is needed to produce x, x is needed for y, and y is needed when a is used. However, even with analysis the small configuration in Circuit H can be folded in numerous ways, into a circuit of one, two, or three components. Even with some pruning a blind transformation strategy is explosive.

7. Conclusion

7.1. Review

This dissertation shows that the discipline of applicative style is a fitting basis for digital hardware design because the abstraction of functionality, upon which applicative style is predicated, is also fundamental to digital design. Functional specifications and digital realizations are given in virtually the same notation. Moreover, the transition of interpretation from instantaneous operation to sequential behavior, *lifting,* is transparent to the basic techniques of this approach. This transparency erases the discontinuity that typically results when design moves from an abstract specification notation to a concrete realization notation.

The design method is to specify an algorithm in a purely functional notation, without regard to representation *or control,* and then to derive from that specification a description of an equivalent digital/synchronous system. I have focused on transformation methods, a form of synthesis in which the engineer is simply "doing algebra" on the formulation of a design. Notation is manipulated by such rules as folding, unfolding, combination, and symbolic simplification, with the goal of reaching a syntactic form that fits the implementation realm.

7.1.1. Iteration.
The secondary notation of a flowchart or finite state machine, which is often used in conventional circuit design, does *not* arise in this method. However, it should be emphasized that this is merely an occlusion of syntax. A major step in each of the examples was to find an *iterative* version of the specification. Iterative form characterizes sequential control (*i.e.* flowchartability); thus, this approach gives the engineer a notation to develop a quality that

is intrinsic to other notations, such as flowcharts. In that sense at least, a functional specification language is more abstract, hence less constraining, than a procedural one. It leaves the way open to develop realizations according to other strategies than the linearization of control.

A *simple loop* can always be constructed from an iterative specification by introducing a parameter that serves as a control token. Simple loops are essentially realizations: Theorem 3.3-5 yields a circuit description immediately by lifting. The elementary functional recursion of the loop transposes to the signal reflexivity of a connectivity description. At the same time, the method admits prevailing structured design techniques. Hierarchical decomposition, through macros (packaged combinations) and representation abstraction (abstract components), are transparent to lifting.

7.1.2. Circuit Synthesis.

A *signal* is a mapping from time to values that subsumes the recurrence relation by which digital systems are usually described. I avoid explicit mention of time by modeling a signal as a sequence and a circuit as a fixed point in the domain of signals. Since behavior is discrete (and since feedback loops always pass through registers), constructing the fixed point is equivalent to inductively solving the corresponding recurrence. This model unifies the mathematical treatment of specification and realization languages and also results in an experimental vehicle for synthesis: DAISY. DAISY's application operator interprets "function-lists" in a manner consistent with (in fact it motivated) the definition of component application in Section 3.5. The choice to model a component as signal of operations—rather than as an operator on signals—is of little consequence in a basic behavioral model because primitive components are constants in that interpretation. However, when circuits are factored into communicating abstract components, the residual *instruction* signal is consistently viewed as a component whose operation varies. The factorization distributes the conditional across application, then distributes application over behavior; as usual, everything lifts.

Experimentation served two purposes in this investigation. It provided the means both to observe circuit behavior and also to certify derivations empirically. In a few instances, observation revealed qualities of performance that are not addressable in the specification language. The discovery in Section 4.4 that the *GCD* circuit stabilizes was an illustration that formal specifications do not account for every quality that a realization might have. The lengthy derivation in Section 5.3 was done entirely by hand, although a number of the steps could be automated using published techniques. A DAISY version of each stage of the derivation was written and executed (Appendix B) on a representative set of inputs. At the very least, this reduced typographical errors, but it also raised the level of our confidence in the derivation. A proof need not be completely correct to be useful (Lipton, *et.al.* 1979); a circuit description must be. Automated synthesis systems are likely always to have gaps that must be bridged empirically, for they free the engineer to think ever more abstractly. The ability to construct and carry out experiments is a significant advantage, if not a necessity, all the more so if it can be done directly in the notation *of* the synthesis system.

7.1.3. Circuit Refinement. Through Chapter 5 the emphasis in synthesis is on manipulation of specifications. If this area is not fully understood, it is at least well charted by research in program synthesis. In passing from specification notation to realization notation the concerns of the designer should become more local, for it is at that point that the monolithic view of the developing description disintegrates, from a simple loop into a system of interconnected but otherwise autonomous components. This in no way implies that all design decisions can be made on the specification-side. As an example of local refinement strategies Chapter 6 presented a "special purpose" transformation system. The specific goal was to use serialization to trade space, measured by the number of external connections, against time. Correct realizations were constructed through a small set of rewriting rules. The derivations introduced registers to implement serialization and therefore also complicated the timing of the circuit. However, they also spawned a *schedule,* for target behavior that could be used to coordinate it with the surrounding system.

7.2. Limitations of the Approach

If one appraises the realization language in terms of typical circuit designs, one can see that it falls short of its fundamental purpose—to "portray implementations" (Chapter 1). The notation makes it difficult to express bidirectionality in signals; whether the difficulty is due to shortcomings in syntax or semantics should be considered carefully. This dissertation only touches on the issue of communication; an important question to consider in judging this approach is how it extends to account for external and independent signals.

144

7.2.1. Bidirectionality.
Since the specification language is purely functional, it is not surprising that an applicative realization language suffices as a target for synthesis. As defined here, circuit descriptions state connectivity using applicative terms that require a distinction between input and output. Consequently, my realization language is inadequate for describing components, such as some memories[1], whose input and output leads are physically identical. Milner overcomes the problem by using a notation for connectivity that does not depend on the input/output distinction (Milner, 1973). The realization language I have adopted translates easily to Milner's notation. What emerges is a relational model of behavior; functionality is a special case. Of course, directionality (perhaps "causality" is a better word) is also obscured in the resultant semantics. Relational specification languages, such as PROLOG (See for example Clark and Gregory, 1981), might be used to confront bidirectionality directly. On the other hand, directionality (functionality) is the preferable abstraction and should not be lightly discarded. One finds evidence for this thesis by looking at how circuit design has evolved away from its natural basis (analog components in equilibrium) to an artificial digital basis that forces a circuit to behave as a function on its state. Bidirectional wires rarely[2] serve simultaneously as both input and output; rather, they are a physical unification of conceptually distinct entities. That point notwithstanding, physically identical parts of an object should surely be identified in the description of that object, and in the case of bidirectional leads this is a problem for the applicative realization language adopted here.

7.2.2. Digital Asynchrony, Communication, and Integration.
My examples all deal with closed specifications and consequently I was able to develop circuit descriptions in a uniform temporal framework. I employed standard techniques to decompose architecture, but said little about decomposition of control. How standard control factorizations (procedures, coroutines, *etc.*) are lifted merits study. When a designer breaks a problem down in this fashion he incurs a liability in the form of a communication problem and must develop a protocol by which autonomous controllers coordinate their activity.

---

[1] See (Mead and Conway, 1980, p. 161, Fig. 5.19) for another fine example.

[2] Counterexamples are wholeheartedly invited.

Digital asynchrony is discrete autonomy. The interval between meaningful external events is an unknown but always integral number of clock cycles. The first law of structured digital design is to "latch" truly asynchronous signals and thereby ensure that, from the point of view of the system, they occur at opportune moments.

The subject of digital asynchrony has been broached several times in this dissertation. For example, one way to introduce autonomous processes is to designate them as operations. That is, assume that they behave in negligible time and deal with coordination separately. In the *L*-circuit of Chapter 5, the *ENVIRONMENT* instructions **EXT**, **FND**, and **LBL**, were presumed to result in trivial operations. However, it is barely credible to assume that *extend, find,* and *label* are trivial[3]. To complete the realization of the *L*-interpreter it will likely be necessary to introduce protocols for waiting, in order to intergrate the autonomous abstract components. In the meantime, a natural strategy for control decomposition is to carry out design-as-usual while treating certain serious symbols as though they were trivial. Some conventional design techniques, for example self-timing strategies (Mead and Conway, 1980), would support this strategy.

A circuit that is party to a communication (and this includes many circuits) cannot be specified in closed form. Its description must account for externally generated signals, and the operator/value based specification language used here must be extended to express input/output. The *single-pulser* discussed by Winkel and Prosser (1980, *pp.* 183-186) is a nice example because its computation is minimal in relation to its communication.

> **Problem Statement.** We have a debounced pushbutton, with the down position meaning on (true) and the up position off (false). Devise a circuit to sense the depression of the button and assert an output signal for one clock pulse. The system should not allow additional assertions of the output until after the operator has released the button.

A solution, below, presumes not only that the button is debounced but also that

---

[3]O'Donnell's associative architecture (1981, 1983) can perform these operations in unit time if some restrictions are made. The question is not whether such things *can* be done, but whether they *will* be done in a particular design endeavor. A conventional implementation, using off-the-shelf components, would certainly require several cycles to implement these operations.

it is latched. The specification for the single-pulser must take into account that some of the identifiers change according to external stimuli. Let us introduce pseudo-operations *get* and *put* that express this. Assume that a depressed button and pulse assertion are both implemented by high voltage. The authors' flowchart specification, expressed as an iterative specification over type **Dig** (Section 2.1), becomes[4]

$$FIND(b,\ p) \Longleftarrow high?(b) \rightarrow WAIT(\ get(b),\ put(high)\ ),$$
$$FIND(\ get(b),\ put(low)\ ).$$

$$WAIT(b,\ p) \Longleftarrow high?(b) \rightarrow WAIT(\ get(b),\ put(low)\ ),$$
$$FIND(\ get(b),\ put(low)\ ).$$

With control token c representing *WAIT/FIND* as *high/low*, a realization is

$$C = \phi\ !\ MUX_2(B,\ C,\ [high\,],\ [low\,],\ [high\,],\ [low\,])$$
$$B = \quad GET(B)$$
$$P = \quad PUT(MUX_2(B,\ C,\ [high\,],\ [low\,],\ [low\,],\ [low\,]))$$

$$where\ mux_2(b,\ c,\ w,\ x,\ y,\ z) \Longleftarrow high?(c) \rightarrow [high?(b) \rightarrow w,\ x],$$
$$[high?(b) \rightarrow y,\ z].$$

Simplification of the conditionals leads to a refined realization:

$$C = \phi\ !\ B$$
$$B = \quad GET(B)$$
$$P = \quad PUT(\ AND(B,\ NOT(C))\ )$$

Put and *get* are coercions from external signal to value; they become redundant when lifted. If they are simply eliminated we arrive at the authors' solution:

---

[4]Since this specification has no base clauses it does not converge. Unless we are careful about the meaning of *put,* its minimal solution is the undefined function.

```
C =  φ !  B
P =        AND(B, NOT(C))
```

Note that since there must be some voltage on every signal at every time the specification reads and writes on every iteration. Admittedly, this is a clumsy way to introduce external communication to the specification language, but at least it is direct. Attempts to isolate a less verbose applicative construct for external communication have lead to a number of proposed constructs for *indeterminacy*. Filman and Friedman survey the variety of approaches in their text (1983). The issue was addressed as early as 1963 by McCarthy, through his AMB operator (1963a). Keller (1978) discusses indeterminacy using Kahn's process semantics as a starting point. There does not appear to be a consensus on that topic at this time. Along the lines of the research reported here, Johnson (1982) shows one way to specify asynchronous systems using the indeterminate constructor of Friedman and Wise (1979, 1980, 1981). This constructor is implemented in DAISY, but was not exploited in this dissertation.

7.3. Prospects for Research

This discussions of the previous section ask basic questions about the foundations of functional style. There are, as well, many refinements to the method presented here that are worthy of investigation.

7.3.1. Multiphase Clocking.
My schematics depict registers as boxes that are governed by a universal clock. The notation and terminology call to mind a printed-wire fabrication medium, where the qualities of a storage *component* are consistent with the pictures. In other media storage elements can be less physically imposing, and can also give rise to other synchronization strategies. In VLSI designs, for example, storage is sometimes implemented with pass transistors and synchronized by alternating clocking signals (Mead and Conway, 1980). *Multiphase clocking* could be expressed in my realization language through a partitioning of storage elements (the term "register" becoming counterintuitive at this point) according to the phases they serve. One might obtain a canonical 2-phase system in a form like "$Z = z^0 ! z^1 ! G(Z)$" and then proceed to make refinements. How a properly phased realization can be synthesized merits study;

and may also be a key to addressing the bidirectionality problem (Section **7.1.1**).

7.3.2. The Realization Language as a Formal System   Aschroft
and Wadge (1976) present LUCID as a formal system in the tradition of Hoare's
(1969). (I noted the similarity between my realization language and LUCID in
Section 1.2.1.) A LUCID specification can be viewed as a set of axioms, used to
deduce assertions about behavior. The works just cited address correctness;
hence, description text is used to generate verification conditions. Although I
have adopted synthesis as a means for dealing with correctness, in the course of
experimentation other kinds of observations were made about circuit behavior.
Aschroft and Wadge point out that LUCID can be used to address other properties, and it would be interesting to explore how the realization language might be
used to generate "performance conditions" about stability, power consumption,
fault tolerance, and so on. Hafner and Parker (1983) do just that; they use a
behavioral description language, syntactically similar to mine, to synthesize timing requirements.

There is also the intriguing possibility that with appropriately redefined base
operations, realizations themselves might construct performance characterizations
or fabrication data. For example, since recursion corresponds directly to connectivity in realizations, a graphics data base could be established by evaluating a
realization in an environment where the ground symbols are bound to graphics
primitives.

7.3.3. Other Topics.   This dissertation gives additional motivation for the
continued study of transformability among recursion schema, and other general
problems surrounding the automation of synthesis. Research is needed not only
to formalize semantics but also to address the nature of interaction in synthesis
systems. If one stipulates a component of human creativity in computer-aided
design, then it is not enough simply to require of the human all that the computer cannot or has not yet been programmed to do.

It was noted in Section 1.1 that design is dualistic: it is characterized as an
interplay between the selection of an algorithm and the selection of a representation in which that algorithm executes. This holds in software and hardware
alike, and this dissertation makes only modest inroads into the problem area of
choosing a representation. This is an open area for research, but the question
that follows from this investigation can be stated simply: "which methods lift?"

7.4. Final Remarks

I prefer the game of GO to the game of CHESS. It stimulates me more, although differently. Since I am a master of neither game, my preference is hardly authoritative; but even if I were a master of both my preference would not make GO a better game.

I did not set out to prove in this dissertation that applicative methods are better than others for the design of circuits. The question I asked myself was whether the constraints of the the style would *allow* one to describe circuits, and if so, are there any advantages in using the style for that purpose. That one can *describe* circuits in a purely applicative way, though perhaps moderately surprising at first, says nothing about the practicality of doing it. However, that one can *derive* a realization by "doing ordinary algebra" indicates that the approach is indeed a promising basis for engineering. This inference depends on the reader's agreement, first, that the target notation achieves its concretely descriptive purpose (I believe that to be self evident); and second, that the specification language is a suitable notation for expressing ideas. The second point is a premise of this work; to conclude here that the approach is superior to conventional methods would be to beg the question. Still, I think that those who are familiar with digital design will, in retrospect, find substantial benefit in applicative style.

For those already predisposed to McCarthy's basis, this dissertation has something further to say about its appropriateness and its relationship to programming. It is additional positive evidence presented in a more neutral (*i.e.* less von Neumann) setting. To compare functional languages to procedural ones is, to a large extent, to compare specifications to their realizations, or for that matter, GO to CHESS. If, nevertheless, one is resolute to make a comparison, it should be done on the basis of an independent target language. Digital systems seem more suitable than, say, machine code for this purpose. I think I have shown applicative methods to be competitive in that realm, and I hope that the evidence herein is sufficient to provoke further investigation. I also hope that CHESS players who would follow the progress of that investigation try a few games of GO.

Selected Bibliography

Ashcroft, Edward A. and William W. Wadge, Lucid, a nonprocedural language with iteration. *Comm. ACM,* **20**(7):519–526 (July, 1977).

Auslander, Marc A. and H.R. Strong, Systematic recursion removal. *Comm. ACM,* **21**(2):127–133 (February, 1978).

Backus, John, Can programming be liberated from the von Neumann style? *Comm. ACM,* **21**(4):613–641 (August, 1978).

Backus, John (1981a), The algebra of functional programming: function level reasoning, linear equations, and extended definitions. *Proc. of the Symposium on Functional Languages and Computer Architecure, eds.* B. Nordstrum, A. Wikstrom, and Soren Holmstrom, Goteborg, Sweden, June, 1981, 408–450.

Backus, John (1981b), Function level programs as mathematical objects. *Proc. of the 1981 ACM Conference on Functional Programming Languages and Computer Architecture,* (ACM order no. 556810), 1–10.

Backus, John, F.L. Baur, J. Green, C. Katz, J. McCarthy, P. Naur, A.J. Perlis, H. Rutihauser, K. Samelson, B. Vauquois, J.J. Wegstein, A. van Wijngaarden, and M. Woodger, Revised report on the algorithmic language ALGOL 60. *Numer. Math.* **4**:420–453 (1963). Also published in *Comm. ACM,* **6**(1):1–17 (January, 1963).

Brainerd, Walter S. and Lawrence H. Landweber, *Theory of Computation,* Wiley and Sons, New York, 1974.

Burstall, Rod M. and John Darlington, A transformation system for developing recursive programs. *J. Assoc. Comput. Mach.,* **24**(1):44–67, (January, 1977).

Burge, William H., *Recursive Programming Techniques,* Addison-Wesley, Reading, Pa., 1975.

Cardelli, Luca, *An Algebraic Approach to Hardware Description and Verification,* Ph. D. dissertation, Univ. of Edinburgh, 1982.

Cardelli, Luca, Analog processes. *Proc. of the Ninth Symposium on Mathematical Foundations of Computer Science, Lecture Notes in Compter Science, No. 88,* Springer, New York, 1980, 181–193.

Chandra, Ashok K., Efficient compilation of linear recursive programs. Stanford Artificial Intelligence Memo AIM-169, Technical Report STAN-CS-282, Dept. of Computer Science, Stanford University, April, 1972.

Cheatham, Thomas E., Jr., Glen H. Holloway, and Judy A. Townley, Symbolic evaluation and the analysis of programs. *IEEE Trans. Software Engrg.*, **SE-5**(4):402–417, (July, 1979).

Clark, Keith L., and Steve Gregory, A relational language for parallel programming. *Proc. of the 1981 ACM Conference on Functional Programming Languages and Computer Architecture*, (ACM order no. 556810), 171–178.

Cohen, Norman Howard, *Source-to-source Improvement of Recursive Programs*, Ph.D. dissertation, Harvard Univ., Cambridge, Mass., 1980.

Cooper, David C., Bohm and Jacopini's reduction of flow charts. *Comm. ACM*, **10**(8):463, 473 (August, 1967).

Cuny, Janice E. and Lawrence Snyder, Conversion from data-flow to synchronous execution in loop programs. Report for the BLUE CHiP Project, Purdue University Department of Computer Sciences, West Lafayette, Indiana, 1982.

Darlington, John and Rod M. Burstall, A System which automatically improves programs. *Acta Informat.*, **6**:41–60, 1976.

De Millo, Richard A., Richard J. Lipton, and Alan J. Perlis, Social Processes and proofs of theorems. *Comm. ACM*, **22**(5):271–280, (May, 1979).

Filman, Robert E. and Daniel P. Friedman, *Coordinated Computing: Tools and Techniques for Distributed Software*, McGraw-Hill, New York, 1983.

Friedman, Daniel P. and David S. Wise, An approach to fair applicative multiprogramming. in *Semantics of Concurrent Computation*, ed. G. Kahn, *Lecture Notes in Compter Science, No. 70*, Springer, New York, 1979 203–226.

Friedman, Daniel P. and David S. Wise, Aspects of applicative programming for file systems. *Proc. ACM Conf. on Language Design for Reliable Software, ACM SIGPLAN Notices*, **12**:41–55, (March, 1977).

Friedman, Daniel P. and David S. Wise (1978a), Aspects of applicative programming for parallel processing. *IEEE Trans. Comput.*, **C-27**(4):289–296, (April, 1978).

Friedman, Daniel P. and David S. Wise (1976a), CONS should not evaluate its arguments. in *Automata, Languages and Programming,* eds. S. Michaelson and R. Milner, Edinburgh Univ. Press, Edinburgh, 1976, 257–284.

Friedman, Daniel P. and David S. Wise, Fancy ferns require little care. *Proc. of the Symposium on Functional Languages and Computer Architecure, eds.* B. Nordstrum, A. Wikstrom, and Soren Holmstrom, Goteborg, Sweden, June, 1981, 124–156.

Friedman, Daniel P. and David S. Wise (1978b), Functional combination. *Computer Languages*, **3**(1):31–35, 1978.

Friedman, Daniel P. and David S. Wise, An indeterminate constructor for applicative programming. *Conf. Rec. 7th ACM Symposium on Principles of Programing Languages*, (January, 1980), 243–250.

Friedman, Daniel P. and David S. Wise (1976b), Output driven interpretation of recursive programs, or writing creates and destroys data structures. *Inform. Process. Lett.*, **5**(6):155–160 (December, 1976); *Erratum:* **9**(2):101 (August, 1979).

Friedman, Daniel P. and David S. Wise (1976c), Unbounded computational structures. *Software - Practice and Experience*, **8**:407–416 (1976).

Friedman, Daniel P., David S. Wise, and Mitchell Wand, Recursive programming through table lookup. *ed.* R.D. Jenks, *Proc. 1976 ACM Symposium on Symbolic and Algebraic Computation*, 85–89.

Gannon, Dennis, Pipelining array computations for MIMD parallelism: a function specification. *Proc. of the 1982 International Conference on Parallel Processing*, IEEE (order no. 421), 1982, 284–286.

Garland, Stephen J. and David C. Luckham, Translating recursion schemes into program schemes. *Proc. of an ACM Conf. on Proving Assertions about Programs*, Las Cruces, New Mexico, January, 1972, published as *SIGPLAN Notices* **7**(1) and *SIGACT News* No. 14., (January, 1972), 83–96.

Gordon, Michael J.C., *The Denotational Description of Programming Languages, An Introduction*, Springer, 1979.

Gordon, Michael J.C., The denotational semantics of sequential machines. *Inform. Process. Lett.*, **10**(1):1–3, (February, 1980).

Gordon Michael J.C. (1981a), A model of register transfer systems with applications to microcode and VLSI correctness. Corrected version of Dept. of Computer Science Internal Report CSR-82-81, Univ. of Edinburgh, 1981.

Gordon Michael J.C. (1981b), A very simple model of sequential behavior of *n* mos. *Proc. of the VLSI 81 International Conference*, Edinburgh, August, 1981.

Greibach, Sheila A., *Theory of Program Structures: Schemes, Semantics, Verification, Lecture Notes in Compter Science, No. 36*, Springer, New York, 1975.

Hafer, Louis J., and Alice C. Parker, Automated synthesis of digital hardware. *IEEE Trans. Comput.*, **C-31**(2):93–109 (February, 1981).

Hafer, Louis J., and Alice C. Parker, A formal method for the specification, analysis, and design of register–transfer level digital logic. *IEEE Trans. on Computer-aided Design of Integrated Circuits and Systems*, **CAD-2**(1):4–18 (January, 1983).

Harel, David, On folk theorems. *Comm. ACM,* **23**(7):379–389 (July, 1980).

Henderson, Peter, *Functional Programming: Application and Implementation*, Prentice-Hall, Englewood Cliffs, 1980.

Henderson, Peter, and James H. Morris, Jr., A lazy evaluator. *Conf. Rec. Third ACM Symposium on Principles of Programming Languages*, 1976, 95–103.

154

Hill, Fredrick J. and Gerald R. Peterson, *Introduction to Switching Theory and Logical Design,* (third ed.), Wiley and Sons, New York, 1968.

Hoare, C.A.R, An axiomatic basis for computer programming. *Comm. ACM,* **12**(10):576–580, 583, (October, 1969).

Hoare, C.A.R., Proof of correctness of a data representation. *Acta Informat.,* **1**:271–281 (1972).

Johnson, Steven D., Applicative programming and digital design. To appear, *Eleventh Annual ACM SIGACT-SIGPLAN Symposium on Principles of Programming Languages,* Salt Lake City, January, 1984.

Johnson, Steven D., Circuits and systems: implementing communication with streams. *Proc. 10th IMACS World Congress on Systems Simulation and Scientific Computation,* Vol. 5, *eds.* W.F. Ames and R. Vichnevetsky, Montreal, August, 1982.

Kahn, Gilles, A preliminary theory for parallel programs. Rapport de Recherche n° 6. IRIA Laboria, (January, 1973).

Kahn, Gilles, and David MacQueen, Coroutines and networks of parallel processes. *IFIP 77,* North-Holland, 1977, 933–938.

Keller, Robert M., Denotational models for parallel programs with indeterminate semantics. in *Formal Description of Programming Concepts, ed.* E.J. Neuhold, (Proc. of the IFIP Working Conference, August, 1977) North-Holland, 1978, 337–366.

Kleene, Stephen C., *Introduction to Metamathematics,* North Holland, New York, 1952.

Kohlstaedt, Anne T., Daisy 1.0 reference manual. Technical Report No. 119, Indiana Univ. Computer Science Dept., Bloomington, Indiana, November, 1981.

Landin, Peter J., A correspondence between ALGOL 60 and Church's lambda notation – part I. *Comm. ACM,* **8**(2):89–101, (February, 1965).

McCarthy, John (1963a), A basis for a mathematical theory of computation. *Computer Programming and Formal Systems, eds.* P. Braffort and D. Hirschberg, North-Holland, Amsterdam, 1963, 33–70.

McCarthy, John, Recursive functions of symbolic expressions and their computation by machine, part I. *Comm. ACM,* **3**(4):184–195 (April, 1960).

McCarthy, John (1963b), Towards a mathematical science of computation. *Proc. of the IFIP Congress '62, ed.* C. M. Popplewell, North-Holland, Amsterdam, 1963, 21–28.

McCarthy, John, P.W. Abrahams, D.J. Edwards, T.P. Hart, and M.I. Levin, *Lisp 1.5 Programmer's Manual,* The MIT Press, Cambridge, Mass., 1973.

Manna, Zohar, *Mathematical Theory of Computation,* McGraw-Hill, New York, 1974.

Manna, Zohar, and Richard J. Waldinger, Synthesis: dreams => programs. *IEEE Trans. Software Engrg.*, **SE-5**(4):294–328 (July, 1979).

Manna, Zohar and Richard J. Waldinger, Towards automatic program synthesis. *Comm. ACM*, **14**(3):151–165, (March, 1971).

Mead, Carver and Lynn Conway, *Introduction to VLSI Systems*, Addison-Wesley, Reading, 1980.

Meyers, Thomas J., *Infinite Structures in Programming Languages*, Ph.D. dissertation, University of Pennsylvania, Philadelphia, 1980.

Milne, George and Robin Milner, Concurrent processes and their syntax. *J. Assoc. Comput. Mach.*, **26**(2):302–321, (April, 1979).

Milne, Robert and Christopher Strachey, *A Theory of Programming Language Semantics*, Chapman and Hall, London, 1976.

Milner, Robin (1980a), *A Calculus of Communicating Systems, Lecture Notes in Compter Science, No. 92*, Springer, New York, 1980.

Milner, Robin, Processes: a mathematical model of computing agents. *Proc. Logic Colloq. '73, eds.* Rose and Shepherdson, North-Holland, 1973.

Milner, Robin (1980b), On relating synchrony and asynchrony. Technical Report No. CSR-75-80, Univ. of Edinburgh, Edinburgh, 1980.

Morris, James H., Jr., and Benjamin Wegbreit, Subgoal induction. *Comm. ACM*, **20**(4):209–222, (April, 1977).

Mycroft, Alan, The theory and practice of transforming call-by-need into call-by-value. *Proc. of the Fourth International Symposium on Programming, ed.* B. Robinet, *Lecture Notes in Compter Science, No. 19*, Springer, New York, 1980, 269–281.

O'Donnell, John, A Systolic Associative LISP Computer Architecture with Incremental Parallel Storage Management, Techical Report No. 81-5, Department of Computer Science, University of Iowa, 1981.

O'Donnell, John, (1983) *personal communication.*

Paterson, Michael S. and Carl E. Hewitt, Comparative schematology. in *Record of Project MAC Conference on Concurrent Systems and Parallel Computation*, Association for Computer Machinery, New York, 119–128, (December, 1970).

Scott, Dana S., Data types as lattices. *SIAM J. Comput.*, **5**(3):522–587, (September, 1976).

Scott, Dana S., Domains for denotational semantics. corrected and expanded version of a paper presented at *ICALP 82*, (July, 1982).

Scott, Dana S., Logic and Programming Languages. *Comm. ACM*, **20**(9):634–641, (September, 1977).

Steele, Guy L., Jr. and Gerald J. Sussman, The revised report on Scheme: a dialect of Lisp. MIT Artificial Intelligence Laboratory Memo 452, January, 1978.

Stoy, Joseph E., *Denotational Semantics: The Scott-Strachey Approach to Programming Language Theory*, MIT Press, Cambridge, 1977.

Strong, H.R., Jr., Translating recursion equations into flow charts. *J. Comput. System Sci.*, **5**(6):254–285, (June, 1971).

Tennent, Robert D., The denotational semantics of programming languages. *Comm. ACM*, **19**(8):437–453 (August, 1976).

Vuillemin, Jean, Correct and optimal implementations of recursion in a simple programming language. *J. Comput. System Sci.*, **9**(3):332–354, (March, 1974).

Wadsworth, Christopher, *Semantics and Pragmatics of Lambda-calculus*, Ph.D. dissertation, Oxford, 1971.

Wand, Mitchell (1980a), Continuation based program transformation strategies. *J. Assoc. Comput. Mach.*, **27**(1):164–180, (January, 1980).

Wand, Mitchell (1982a), Deriving target code as a representation of continuation semantics. *ACM Trans. Programming Languages and Systems* **4**(3):496–517

Wand, Mitchell (1980b), Different advice on structuring compilers and proving them correct. Technical Report No. 95, Computer Science Department, Indiana University, Bloomington, September, 1980.

Wand, Mitchell, *Induction, Recursion, and Programming*, North Holland, New York, 1980.

Wand, Mitchell, Loops in Combinator-Based Compilers. *Conf. Rec. 10th ACM Symp. on Principles of Programming Languages* 1983, 190–196.

Wand, Mitchell (1982b), Semantics-directed machine architecture, *Conf. Rec. 9th ACM Symp. on Principles of Programming Languages* (1982), 234–241.

Wand, Mitchell, and Daniel P. Friedman, Compiling Lambda-expressions using continuations and factorizations. *Computer Languages*, **3**:241–263, (1978).

Winkel, David and Franklin Prosser, *The Art of Digital Design*, Prentice-Hall, Englewood Cliffs, New Jersey, 1980.

Wise, David S., Interpreters for functional programming. *Functional Programming and its Applications, eds.* J. Darlington, P. Henderson, and D.A. Turner, Cambridge University Press, Cambridge, 1982, 186–195.

Wise, David S., A powerdomain semantics for indeterminism. Indiana Univ. Technical Report No. 142, July, 1983.

Appendix

A. True Syntax of DAISY

At the time of this writing a parser for DAISY's proposed syntax (Figure 4.1) has not been fully implemented. This appendix gives the present version of the language. Further documentation can be found in Kohlstaedt's programmer's manual (1982), which also cites published research that inspired development of the language. Appendix B shows the DAISY source actually used for examples in this dissertation. The present syntax of the language is given in Figure A-1. Examples of conversions between present syntax and proposed notation are shown in Figure A.2. The primitive conditional unnecessary because the list constructor is non-strict. There is a 3-place operation, **if**, that selects an alternative according to its first argument. The Boolean coercion function *predicate* in Figure 4-4b describes the implementation of **if** accurately. Recursive and lexically scoped systems are built by pseudo operations **rec** and **let**.

$expression$ ::= **❶** $expression$ | $atom$ | $fern$ | $application$ | $abstraction$

$atom$ ::= $identifier$ | $numeral$ | $operator$

$fern$ ::= ($list$) | < $list$ > | { $list$ }
$list$ ::= Λ | $expression$ * | $expression$! $expression$ | $expression$ $list$

$application$ ::= $expression$: $expression$

$abstraction$ ::= \($expression$. $expression$)

$definition$ ::= $expression$ = $expression$
= $identifier$: $expression$ =: $expression$.

Figure A.1. Present DAISY Syntax

DAISY	stylized text
o i	@ i
(x y)	[x y]
(x ! y)	[x ! y]
(x *)	[x *]
<a b>	< a b >
<a ! b>	<a ! b>
<a *>	<a *>
{a b}	{ a b }
{a ! b}	{a ! b}
f : a	f : a
\\(x . e)	λ x . e
if:< p a q b c>	p $\rightarrow$ a, q $\rightarrow$ b, c
let:(x a e)	**let** x = a **in** e
let:((x y) <a b> e)	**let** x = a y = b **in** e
rec:(x a e)	**rec** e **where** x = a
f:x =: e.	F:x $\Longleftarrow$ e.

Figure A.2. Conversions to Present DAISY Syntax.

A *fern* is a "list specification", the salient properties of a list being its content and its order. The three fern delimiters express progressively weaker stipulations about them. The delimiters '(...)' have been changed to '[...]' in the idealized language because parentheses will eventually be reserved for parser direction. Ferns of the form '[...]' denote *pure lists*, a form of structural quotation stating content and order literally. Ferns of the form '<...>' denote *value lists*, whose content depends on the current environment, but whose order is fixed. Value lists act like LISP's LIST operation. Ferns of the form '{...}' denote lists of values, but do not specify an order. The construct is used to address indeterminacy.

Comments in DAISY programs are delimited on the left by a vertical bar, '|', and on the right be a carriage return. Comment lines are used to mimic the proposed notation. For example, the *factorial* realization

$$X = \quad x^0 \,! \; DCR(X)$$
$$Y = \quad 1\,! \; Z$$
$$Z = \quad 1\,! \; MPY(Y, Z)$$
$$READY = \quad ZERO?(X)$$

is implemented in idealized DAISY as

```
FIB:x <= rec test:<X Y READY>
         where
             X = <x ! DCR:<X>>
             Y = <1 ! Z>
             Z = <1 ! ADD:<Y Z>>
         READY =      ZERO?:<A>.
```

The idealized source for the experiment is

```
FIB:x =: rec:((X Y Z READY)
            < <x ! DCR:<X>>
              <1 ! Z>
              <1 ! ADD:<Y Z>>
                  ZERO?:<X>    >
           test:<X Y READY> )
```

With mimicing comments added the source file used was

```
FIB:x =: rec:((X Y Z READY)
          <| X =
                <x ! DCR:<X>>
           | Y =
                <1 ! Z>
           | Z =
                <1 ! ADD:<Y Z>>
           | READY =
                    ZERO?:<X>
          >| in
                test:<X Y READY> )
```

B. DAISY Trials

This appendix contains listings of the Daisy source for experiments of Section 4.4.4 and listings for experiments with the L-interpreter derivation in Chapter 5. Appendix A gives the conversion between the idealized version of Daisy used in the body of this dissertation and the present syntax of the language as reflected here.

The program source listings were printed from the source files used for experimentation. The execution listings were recorded from the actual trial runs, but have been manually modified to clarify the output. In some listings, blank lines were deleted for vertical compression and blank spaces were added to align columns. Repetitive setup commands and responses were deleted from the execution record. Other modifications of the listings are noted where they occur. Included in this appendix are:

- implementations of frequently used components and experimentation aids;

- realizations of the iterative specifications for the *factorial, Fibonacci,* and *greatest-common-divisor* functions, discussed in Section 4.4;

- the realization of the stacking version of the *Fibonacci* specification, discussed in Section 5.1;

- the specifications and realizations generated in the derivation of the L-interpreter circuit in Section 5.3;

- trial forms that were used to test the evolving L-interpreter descriptions; and

- experiments with the L-realization.

Each listing is accompanied by a brief explanation including references to relevant figures and discussions in the body of the dissertation.

```
ADD   = (add*).
DCR   = (\((x).dcr:x)*).
DIV   = (div*).
EQ?   = (eq?*).
IF    = (if*).
INC   = (\((x).inc:x)*).
LT?   = (lt?*).
MPY   = (mpy*).
SUB   = (sub*).
ZERO? = (\((x).eq?:<x 0>)*).
AND   = (and*).
```

Daisy Components. Discussion: Section 4.4.1. Compare with Figure 4.6.

```
test:x =: format:transpose:x.        | Print signals in parallel
                                     |
transpose = (\(x.x)*).               |  - time slices
                                     |
format:(c!S) =: <CR c ! format:S>.   |  - iterleave carraige contol
                                     |
CR = 1:parse:(()).                   |  - carraige control character
```

Experimental Aids. Discussion: Section 4.4.3. The assignment for **CR** is a way to obtain the carraign-return character (ASCII **0D**, hexadecimal), which is not available by name in Daisy.

```
FAC:x =: rec:((X Y READY)
        <| X =
                <x ! DCR:<X>>
         | Y =
                <1 ! MPY:<X Y>>
         | READY =
                ZERO?:<X>
        >| in
                test:<X Y READY> ).


FIB:x =: rec:((X Y Z READY)
        <| X =
                <x ! DCR:<X>>
         | Y =
                <1 ! Z>
         | Z =
                <1 ! ADD:<Y Z>>
         | READY =
                ZERO?:<X>
        >| in
                test:<X Y READY> ).


GCD:(x y) =: rec:((X Y U W V READY)
          <| X =
                  <x ! U>
           | Y =
                  <y ! SUB:<W U>>
           | U =
                  IF:<V X Y>
           | W =
                  IF:<V Y X>
           | V =
                  LT?:<X Y>
           | READY =
                  EQ?:<X Y>
          >| in
                  test:<X Y READY> ).
```

Daisy Source for the Example Realizations. This is source for experimentation with the realizations of the iterative specifications for *factorial* (Figure 4.7),
Fibonacci (Figure 4.8), and *greatest common divisor* (Figure 4.9). Executions of
these descriptions are shown in the figures.

```
evlst:parse:dski:@'/usiu/sdj/PhD/thesis/complib
evlst:parse:dski:@'/usiu/sdj/PhD/thesis/tools
                                          | List representation for stacks.
MTstk = (?? *).                           | Empty stack
                                          | Stack operations:
empty?:(s) =: same?:<s MTstk>.            |    empty:STACK --> BOOL
top:((t ! s)) =: t.                       |     top:STACK --> VALUE
noop:s =: s.                              |    noop:STACK --> STACK
pop:(t ! s) =: s.                         |     pop:STACK --> STACK
push:(v s) =: <v ! s>.                    |    push:VALUE x STACK --> STACK
plop:(v (t ! s)) =: <v ! s>.              |    push:VALUE x STACK --> STACK
                                          |
operate:(i v s) =:                        | Instruction decoder.
   if:< same?:<i @noop> s                 |
        same?:<i @pop > pop:s             |
        same?:<i @push> push:<v s>        |
        same?:<i @plop> plop:<v s>>.      |
                                          | Higher Level Stack Component.
STACK:(s0 I V) =: rec:(S                  |
   | S =                                  |   STACK:[s0 I V] <=
        <s0 ! <operate*>:<I V S>>         |     rec <TOP:S EMPTY?:S>
   | in                                   |     where S = s0 ! OPERATE:<I V S>.
        <<top*>:<S> <empty?*>:<S>>).      |
```

Stack Representation. Stacks are represented as lists for the stacking realization of the *Fibonacci* specification in Section 5.1. There is a discussion of the abstract component *STACK* toward the end of that section. Idealized source for the experiment is given in Figure 5.2a; and the experiment itself is shown in Figure 5.2b. The first two lines obtain component definitions and experimental aids from files named "complib" and "tools." Read the atom '??' as ϕ. The empty stack is an infinite list of don't-knows. Definitions of abstract operations are straightforward (*e.g.* **push** is **cons**), except perhaps for **top**, which has additional formal argument structure because it will be used in a component (see the discussion in Section 4.4.1). The function **operate** serves as an instruction decoder in the abstract component.

```
MUX-N = (mux-N*).                          | Higher Level Multiplexor
mux-N:(p q r u v w x) =:                    |
   if:<p if:<q u v> if:<r w x>>.            |
                                            | Stacking version of Fibonacci
FIBckt:(l0 x0 s0 t0) =:                     |
   rec:((L X (V1 E1) (V2 E2) I P Q READY)
      <|L=
            <l0 ! MUX-N:<P Q V2 <1*> <0*> <1*> <0*>> >
       |X=
            <x0 ! MUX-N:<P Q V2 <1*> DCR:<DCR:<X>> ADD:<X V1> V1> >
       |[V1 E1]=
               STACK:<s0 I MUX-N:<P Q V2 ('#*) DCR:<X> ('#*) X>>
       |[V2 E2]=
               STACK:<t0 I MUX-N:<P Q V2 ('#*) <<>*> ('#*) <@tt*>>>
       |I=
               MUX-N:<P Q V2 (noop*) (push*) (pop*) (plop*)>
       |P=
               EQ?:<L <0*>>
       |Q=
               LT?:<X <2*>>
       |READY=
               AND:<EQ?:<L <1*>> E2>
      >|in
        test:<READY X I V1 L V2 E2 P Q> ).

fib:n =: FIBckt:<0 n MTstk MTstk>.
```

Stacking Realization of FIB. The packaged component **MUX-N** is called *MUX$_4$* in Section 5.1. The help function **fib** initializes registers for experiments. Execution of this description is shown in Figure 5.2b.

```
num?:(tg lf rt) =: same?:<tg @NUM>.        |L-machine type predicates
ide?:(tg lf rt) =: same?:<tg @IDE>.        | Expression types.
lam?:(tg lf rt) =: same?:<tg @LAM>.        |    Numeral
lbl?:(tg lf rt) =: same?:<tg @LBL>.        |    Identifier
apl?:(tg lf rt) =: same?:<tg @APL>.        |    Lambda-exp, \Ide.Exp
cnd?:(tg lf rt) =: same?:<tg @CND>.        |    Reflexive-exp, Ide <= Exp
tst?:(tg lf rt) =: same?:<tg @TST>.        |    Application, Exp : Exp
                                           |    Conditional, Exp -> Exp,Exp
                                           |       rgt-part of conditional
                                           |
                                           | Value predicates
bit?:(tg lf rt) =: same?:<tg @BIT>.        |    Boolean
err?:(tg lf rt) =: same?:<tg @ERR>.        |    Error message
opr?:(tg lf rt) =: same?:<tg @OPR>.        |    Operator
ftn?:(tg lf rt) =: same?:<tg @FTN>.        |    Function closure
fix?:(tg lf rt) =: same?:<tg @FIX>.        |    Expression closure
                                           |
                                           | Action types
arg?:(tg lf rt) =: same?:<tg @ARG>.        |    save function, evaluate arg
act?:(tg lf rt) =: same?:<tg @ACT>.        |    apply function
hlt?:(tg lf rt) =: same?:<tg @HLT>.        |    halt
                                           |
                                           | Constructors and extractors
make-FTN:closure =: <@FTN ! closure>.      |    function closure
make-ERR:message =: <@ERR message>.        |    error message
make-ACT:action =: <@ACT action>.          |    apply action
make-ARG:argument =: <@ARG argument>.      |    evaluate-argument action
tag:((t l r e)) =: t.                      |
lft:((t l r e)) =: l.                      |
rgt:((t l r e)) =: r.                      |
cls:((t l r e)) =: e.                      |
                                           |
                                           | Constants
halt = <@HLT>.                             |    initial continuation
??? = <@???*>.                             |    don't-care signal
```

Representation of the Underlying Type for the L-interpreter. Discussion: Section 5.3.2. All concrete types are represented as lists. Continued on the following two pages.

```
test:(p c a) =:                      | Primitive Conditional
   let:((p+tag p+val) p              |
   if:< bit?:p if:<p+val c a> <>>).  |
                                     |
                                     | Primitive application
apply:((tag ! op) opnd) =: op:opnd.  |
                                     | Stack operations
push:(actn env stk) =: <actn env ! stk>. |
pop:(actn env ! stk) =: stk.         |
top:(actn env ! stk) =: <actn env>.  |
                                     | Components
IF = <if*>.                          |
AND = <and*>.                        |
ACT? = <\((x).act?:x)*>.             |
transpose:x =: (id*):x.              |
id:x =: x.                           |
TOP = <\((x).top:x)*>                |   component version of top
POP = <\((x).pop:x)*>                |   component version of pop
MAKE-ACT = <\((x).make-ACT:x)*>      |   component version of make-ACT
MAKE-ARG = <\((x).make-ARG:x)*>      |   component version of make-ARG
                                     |
                                     | Main multiplexor
                                     |------------------------------
slct:(ctl exp e-num e-opr e-ide e-lam
             e-lbl e-apl e-cnd e-ftn
             e-fix e-tst e-arg e-act
             e-err a-opr a-ftn a-err) =:
   if:<same?:<ctl @EVL>
       if:< num?:exp e-num opr?:exp e-opr ide?:exp e-ide lam?:exp e-lam
            lbl?:exp e-lbl apl?:exp e-apl cnd?:exp e-cnd ftn?:exp e-ftn
            fix?:exp e-fix tst?:exp e-tst arg?:exp e-arg act?:exp e-act
            err?:exp e-err >
       same?:<ctl @APL>
       if:< opr?:exp a-opr ftn?:exp a-ftn a-err >>.
SLCT = (slct*).
```

Representation of the Underlying Type for the L-interpreter (cont'd).
SLCT is the main multiplexor for the realization (*cf.* Fig. 5.8).

```
find:((tag ide) env) =: env:ide.              | Environment operations
                                              |
                                              |
extend:(ide val env) =:                       |
   \(x. if:<same?:<x ide> val env:x>).        |
                                              |
                                              |
label:(ide exp env) =:                        |
   rec:(rho                                   |
        extend:<ide <OFIX exp rho> env>       |
        rho).                                 |
                                              | Initial environment
                                              |--------------------------------
initenv = let:( error make-ERR:(nonnumeric operand)
\(i. if:<
    same?:<i Ozed?>  <OOPR ! \(v. if:<num?:v <OBIT eq?:<2:v 0>> error>) >
    same?:<i Oone?>  <OOPR ! \(v. if:<num?:v <OBIT eq?:<2:v 1>> error>) >
    same?:<i Oinc >  <OOPR ! \(v. if:<num?:v <ONUM inc:2:v> error>) >
    same?:<i Odcr >  <OOPR ! \(v. if:<num?:v <ONUM dcr:2:v> error>) >
    same?:<i Olt? >  <OOPR ! \(u. if:<num?:u
                     <OOPR ! \(v. if:<num?:v <OBIT lt?:<2:u 2:v>> error>)>
                                   error >) >
    same?:<i Oeq? >  <OOPR ! \(u. if:<num?:u
                     <OOPR ! \(v. if:<num?:v <OBIT eq?:<2:u 2:v>> error>)>
                                   error >) >
    same?:<i Oadd >  <OOPR ! \(u. if:<num?:u
                     <OOPR ! \(v. if:<num?:v <ONUM add:<2:u 2:v>> error>)>
                                   error >) >
    same?:<i Osub >  <OOPR ! \(u. if:<num?:u
                     <OOPR ! \(v. if:<num?:v <ONUM sub:<2:u 2:v>> error>)>
                                   error >) >
    same?:<i Ompy >  <OOPR ! \(u. if:<num?:u
                     <OOPR ! \(v. if:<num?:v <ONUM mpy:<2:u 2:v>> error>)>
                                   error >) >
    make-ERR:<OUNBOUND i>                                     >) ).
```

Representation of the Underlying Type for the L-interpreter (cont'd).
Initenv is a function that initially maps operator symbols to operations. Operations are function closures, tagged as type **OPR**. Binary operations for *L* are implemented as *curried* versions of Daisy's operations.

```
M:(exp env) =:
   let:((tag lft rgt) exp
   if:<
       num?:exp   exp
       ide?:exp   COERCE:find:<exp env>
       lam?:exp   make-FTN:<lft rgt env>
       lbl?:exp   M:<rgt label:<lft rgt env>>
       apl?:exp   APPLY:< M:<lft env> M:<rgt env>>
       cnd?:exp   let:((rgt+tag rgt+lft rgt+rgt) rgt
                  test:< M:<lft env> M:<rgt+lft env> M:<rgt+rgt env> >)
       >).

COERCE:val =: if:<
                  opr?:val   val
                  num?:val   val
                  err?:val   val
                  ftn?:val   val
                  fix?:val   let:((val+tag val+exp val+env) val
                                  M:<val+exp val+env>)
                  >.

APPLY:(ftn arg) =:
   if:<
       opr?:ftn   apply:<ftn arg>
       ftn?:ftn   let:((ftn+tag ftn+ide ftn+exp ftn+env) ftn
                       M:<ftn+exp extend:<ftn+ide arg ftn+env>> )
       make-ERR:(invalid function)
       >.

try:exp =: M:<exp initenv>.
```

Concrete Non-linear Specification for the L-interpreter. This specification
was derived in Section 5.3-2, and appears in Figure 5.4. This and all of the following specifications are accompanied by a help function **try** that properly initializes the state for "top level" evaluation.

```
M:(exp stk env) =:
   let:((tag lft rgt) exp
   if:<
       num?:exp   RETURN:<exp stk>
       ide?:exp   COERCE:<find:<exp env> stk>
       lam?:exp   RETURN:<make-FTN:<lft rgt env> stk>
       lbl?:exp   M:<rgt stk label:<lft rgt env>>
       apl?:exp   M:<lft push:<make-ARG:rgt env stk> env>
       cnd?:exp   M:<lft push:<rgt env stk> env>
       >).

COERCE:(val stk) =:
   let:((tag exp env) val
   if:<
       opr?:val   RETURN:<val stk>
       num?:val   RETURN:<val stk>
       err?:val   RETURN:<val stk>
       ftn?:val   RETURN:<val stk>
       fix?:val   M:<exp stk env>
       >).

RETURN:(val stk) =:
   let:((nxt env)      top:stk
   let:((tag lft rgt) nxt
   let:( stk          pop:stk
   if:<
       tst?:nxt   M:<test:<val lft rgt> stk env>
       arg?:nxt   M:<lft push:<make-ACT:val <> stk> env>
       act?:nxt   APPLY:<lft val stk>
       hlt?:nxt   val
       >))).

APPLY:(ftn arg stk) =:
   let:((tag ide exp env) ftn
   if:<
       opr?:ftn   RETURN:<apply:<ftn arg> stk>
       ftn?:ftn   M:<exp stk extend:<ide arg env>>
       RETURN:<make-ERR:(invalid function) stk>
       >).

try:exp =: M:<exp <halt> initenv>.
```

Stacking Version of the L-interpreter. Discussion: Section 5.3.4
(*cf.* Fig. 5.5).

```
M:(ctl ftn arg val exp stk env) =:
   let:((f+tag f+ide f+exp f+env) ftn
   let:((v+tag v+exp v+env) val
   let:((e+tag lft rgt) exp
   let:((nxt old) top:stk
   let:((n+tag n+lft n+rgt) nxt
   let:(stk' pop:stk
   if:<
      same?:<ctl 0EVL>
      if:<
         num?:exp  M:<0RTN <> <> exp <> stk env>
         ide?:exp  M:<0CRC <> <> find:<exp env> <> stk env>
         lam?:exp  M:<0RTN <> <> make-FTN:<lft rgt env> <> stk env>
         lbl?:exp  M:<0EVL <> <> <> rgt stk label:<lft rgt env>>
         apl?:exp  M:<0EVL <> <> <> lft push:<make-ARG:rgt env stk> env>
         cnd?:exp  M:<0EVL <> <> <> lft push:<rgt env stk> env>
         >
      same?:<ctl 0CRC>
      if:<
         err?:val  M:<0RTN <> <> val <> stk env>
         opr?:val  M:<0RTN <> <> val <> stk env>
         num?:val  M:<0RTN <> <> val <> stk env>
         ftn?:val  M:<0RTN <> <> val <> stk env>
         fix?:val  M:<0EVL <> <> <> v+exp stk v+env>
         >
      same?:<ctl 0RTN>
      if:<
         hlt?:nxt  val
         tst?:nxt  M:<0EVL <> <> <> test:<val n+lft n+rgt> stk' old>
         arg?:nxt  M:<0EVL <> <> <> n+lft push:<make-ACT:val <> stk'> old>
         act?:nxt  M:<0APL n+lft val <> <> stk' old>
         >
      same?:<ctl 0APL>
      if:<
         opr?:ftn  M:<0RTN <> <> apply:<ftn arg> <> stk env>
         ftn?:ftn  M:<0EVL <> <> <> f+exp stk extend:<f+ide arg f+env>>
                   M:<0RTN <> <> make-ERR:(invalid function) <> stk env>
         >
      >))))))).

try:exp =: M:<0EVL <> <> <> exp push:<halt <> <>> initenv>.
```

First Loop Version of the L-interpreter. Discussion: Section 5.3.5
(*cf.* Fig. 5.6).

```
M:(ctl val exp stk env) =:
   let:((tag lft rgt env') exp
   let:((nxt old) top:stk
   let:(stk' pop:stk

   if:<same?:<ctl 0EVAL>
      if:<
         hlt?:exp  val
         num?:exp  M:<0EVAL  exp nxt stk' old>
         opr?:exp  M:<0EVAL  exp nxt stk' old>
         ide?:exp  M:<0EVAL  <> find:<exp env> stk env>
         lam?:exp  M:<0EVAL  make-FTN:<lft rgt env> nxt stk' old>
         lbl?:exp  M:<0EVAL  <> rgt stk label:<lft rgt env>>
         apl?:exp  M:<0EVAL  <> lft push:< make-ARG:rgt env stk> env>
         cnd?:exp  M:<0EVAL  <> lft push:<rgt env stk> env>
         ftn?:exp  M:<0EVAL  exp nxt stk' old>
         fix?:exp  M:<0EVAL  <> lft stk rgt>
         tst?:exp  M:<0EVAL  <> test:<val lft rgt> stk env>
         arg?:exp  M:<0EVAL  <> lft push:< make-ACT:val <> stk> env>
         act?:exp  M:<0APPLY val lft stk env>
         err?:exp  M:<0EVAL  exp nxt stk' old>
         >
      same?:<ctl 0APPLY>
      if:<
         opr?:exp  M:<0EVAL  apply:<exp val> nxt stk' old>
         ftn?:exp  M:<0EVAL  <> rgt stk extend:<lft val env'>>
         M:<0EVAL make-ERR:(invalid function) nxt stk' old>
         >
      >))).

try:exp =: M:<0EVAL <> exp push:<halt <> <>> initenv>.
```

Refined Loop for the L-machine. Discussion: Section 5.3.6 (*cf.* Fig. 5.7).

```
ENVIRONMENT:(INSTR ARG-1 ARG-2 ARG-3) =:           |
rec:((FND ENV)                                     |
 <| FND =                                          |
       <find*>:<ARG-1 ENV>                          |
  | ENV =                                          |
       <initenv ! SLCT-E:<INSTR ENV               |
                                ARG-1
                                <label*>:<ARG-1 ARG-2 ENV>
                                <extend*>:<ARG-1 ARG-2 ARG-3>>>
 >| in                                             |
    <FND ENV>).

SLCT-E = <slct-e*>.                                | ENVIRONMENT's
slct-e:(i v0 v1 v2 v3) =:                          | instruction decoder.
   if:< same?:<i @hld>  v0                         |
       same?:<i @set>   v1                         |
       same?:<i @fix>   v2                         |
       same?:<i @ext>   v3 >.                      |
                                                   |
                                                   |
                                                   |
STACK:(INSTR ARG-1 ARG-2) =:                       |
rec:((STK (NXT OLD))                               |
 <|  STK =                                         |
    <<halt <>> ! SLCT-S:<INSTR STK <push*>:<ARG-1 ARG-2 STK> POP:<STK>>>
  |  [NXT OLD] =
     transpose:TOP:<STK>                           |
 >| in                                             |
       <NXT OLD>).                                 |
                                                   |
SLCT-S = <slct-s*>.                                | STACK's
slct-s:(i v0 v1 v2) =:                             | instruction decoder.
   if:< same?:<i @nop> v0                          |
       same?:<i @psh> v1                           |
       same?:<i @pop> v2 >.                        |
```

Abstract Components for the Realizations. Discussion: Section 5.3.8 (*cf.* Fig. 5.8).

```
M:(ctl0 val0 exp0 stk0 env0) =:
   rec:((CTL EVL APL                        | Control register
         VAL ALU FTN ERR                    | Value register
         EXP FND TAG LFT RGT SAV TST        | Expression reg.
         STK (NXT OLD) STK' PSH ACTN ARG ACT | Stack register
         ENV FIX EXT   )                    | Environment reg.

<| CTL =
         <ctl0 ! SLCT:< CTL EXP EVL EVL EVL EVL EVL EVL EVL EVL
                            EVL EVL EVL APL EVL EVL EVL EVL>>
 | EVL =
         <0EVL*>
 | APL =
         <0APL*>
 | VAL =
         <val0 ! SLCT:< CTL EXP EXP EXP ??? FTN ??? ??? ??? EXP
                            ??? ??? ??? VAL EXP ALU ??? ERR >>
 | ALU =
             <apply*>:<EXP VAL>
 | FTN =
             <make-FTN*>:<LFT RGT ENV>
 | ERR =
             <make-ERR:(invalid function)*>

 | EXP =
         <exp0 ! SLCT:< CTL EXP NXT NXT FND NXT RGT LFT LFT NXT
                            LFT TST LFT LFT NXT NXT RGT NXT >>
 | FND =
             <find*>:<EXP ENV>
 | TAG =
             <tag*>:<EXP>
 | LFT =
             <lft*>:<EXP>
 | RGT =
             <rgt*>:<EXP>
 | SAV =
             <cls*>:<EXP>
 | TST =
             <test*>:<VAL LFT RGT>
 | -=> (continued)
```

First L-realization. Discussion: Section 5.3.7 (*cf.* Fig. 5.9). Continued, next page.

```
|     (continued) <=-
|
| STK =
        <stk0 ! SLCT:< CTL EXP STK' STK' STK  STK' STK  PSH  PSH  STK'
                          STK  STK  PSH  STK  STK' STK' STK  STK' >>
| [NXT OLD] =
               transpose:TOP:<STK>
| STK' =
               POP:<STK>
| PSH =
               <push*>:<ACTN ENV STK>
| ACTN =
               SLCT:<CTL EXP ??? ??? ??? ??? ??? ARG RGT ???
                          ??? ??? ACT ??? ??? ??? ??? ??? >
| ARG =
               MAKE-ARG:<RGT>
| ACT =
               MAKE-ACT:<VAL>
|
| ENV =
        <env0 ! SLCT:< CTL EXP OLD OLD ENV OLD FIX ENV ENV OLD
                          RGT ENV ENV ENV OLD OLD EXT OLD >>
| FIX =
               <label*>:<LFT RGT ENV>
| EXT =
               <extend*>:<LFT VAL SAV>
>|in
   monitor:<CTL VAL EXP NXT>).

monitor:((la!ld)(va!vd)(ea!ed)(ta!td)) =:
    <la va ea ta cr ! if:<hlt?:ea <> monitor:<ld vd ed td>>>.
try:exp =: M:<OEVL <> exp push:<halt <> <>> initenv>.
```

First L-realization (cont'd). The help function **monitor** traces registers CTL, VAL, EXP, and NXT, and terminates the trace if the *halt*-action shows up in EXP.

```
M:(ctl0 val0 exp0) =:
    rec:((CTL                                | state label
          VAL ALU FTN ERR                    | values
          EXP TAG LFT RGT SAV TST            | expression
          (NXT OLD) S1 S2 ARG ACT            | stack
          (FND ENV) EO  E1  E2 )             | environment
  |
 <| CTL =
          <ctl0 ! IF:<ACT?:<EXP> <QAPL*> <QEVL*>>>
  | VAL =
          <val0 ! SLCT:< CTL EXP EXP EXP ??? FTN ??? ??? ??? EXP
                                    ??? ??? ??? VAL EXP ALU ??? ERR >>
  | ALU =
                  <apply*>:<EXP VAL>
  | FTN =
                  <make-FTN*>:<LFT RGT ENV>
  | ERR =
                  <make-ERR:(invalid function)*>

  | EXP =
          <exp0 ! SLCT:< CTL EXP NXT NXT FND NXT RGT LFT LFT NXT
                                    LFT TST LFT LFT NXT NXT RGT NXT >>
  | TAG =
                  <tag*>:<EXP>
  | LFT =
                  <lft*>:<EXP>
  | RGT =
                  <rgt*>:<EXP>
  | SAV =
                  <cls*>:<EXP>
  | TST =
                  <test*>:<VAL LFT RGT>
  |
  | -=> (continued)
```

Refined L-realization. Discussion: Section 5.3.8 (*cf.* Fig. 5.10). Continued, next page.

```
|     (continued)<=-
|
|[NXT OLD] =
                    STACK:<S1 S2 ENV>
|  S1 =
                    SLCT:< CTL EXP <0pop*> <0pop*> <0nop*> <0pop*>
                                   <0nop*> <0psh*> <0psh*> <0pop*>
                                   <0nop*> <0nop*> <0psh*> <0nop*>
                                   <0pop*> <0pop*> <0nop*> <0pop*> >
|  S2 =
                    SLCT:<CTL EXP ??? ??? ??? ??? ??? ARG RGT ???
                                  ??? ??? ACT ??? ??? ??? ??? ??? >
|  ARG =
                    MAKE-ARG:<RGT>
|  ACT =
                    MAKE-ACT:<VAL>
|
|  [FND ENV] =
                    ENVIRONMENT:<E0 E1 E2 SAV>
|  E0 =
                    SLCT:<CTL EXP <0set*> <0set*> <0hld*> <0set*>
                                  <0fix*> <0hld*> <0hld*> <0set*>
                                  <0set*> <0hld*> <0hld*> <0hld*>
                                  <0set*> <0set*> <0ext*> <0set*>>
|  E1 =
                    SLCT:<CTL EXP OLD OLD EXP OLD LFT ??? ??? OLD
                                  RGT ??? ??? ??? OLD OLD LFT OLD >
|  E2 =
                    SLCT:<CTL EXP ??? ??? ??? ??? RGT ??? ??? ???
                                  ??? ??? ??? ??? ??? ??? VAL ??? >
>|in
   monitor:<CTL VAL EXP NXT>).

monitor:((ca!cd)(va!vd)(ea!ed)(na!nd)) =:
   <ca va ea na cr ! if:<hlt?:ea <> monitor:<cd vd ed nd>>>.

try:exp =: M:<0EVL <> exp>.
```

Refined L-realization (cont'd). The help function **monitor** traces registers CTL, VAL, EXP, and NXT, and terminates the trace if the halt-action shows up in EXP.

```
    |
    | form - 5
    |
tst1 = (NUM 5).

    |
    | form - (\i.i):5
    |
tst2 = (APL (LAM I (IDE I)) (NUM 5)).

    |
    | form - one?:0 -> err, zero?:0 -> 0, err
    |
tst3 = (CND (APL (IDE one?) (NUM 0)) (TST (ERR 0)
                                        (CND (APL (IDE zed?) (NUM 0))
                                            (TST (NUM 0) (ERR 1)))))).

    |
    | form - (((\f.(\a. f:a)):(\x.inc:x)):5
    |
tst4 = (APL (APL (LAM F (LAM A (APL (IDE F) (IDE A))))
                (LAM X (APL (IDE inc) (IDE X))) )
            (NUM 5)).

    |
    | form - F <= \X.\Y. (eq?:X):Y -> X,
    |                    (lt?:X):Y -> (F:X):((sub:Y):X),
    |                    (F:Y):((sub:X):Y).
tstI = (LBL F
        (LAM X
        (LAM Y
        (CND (APL (APL (IDE eq?) (IDE X)) (IDE Y))
            (TST (IDE X)
                (CND (APL (APL (IDE lt?) (IDE X)) (IDE Y))
                    (TST (APL (APL (IDE F) (IDE X))
                                (APL (APL (IDE sub) (IDE Y)) (IDE X)))
                        (APL (APL (IDE F) (IDE Y))
                                (APL (APL (IDE sub) (IDE X)) (IDE Y)))
                ))))))).

gcd:(x y) =: try: <@APL <@APL tstI <@NUM x>> <@NUM y>>.
```

Trial Forms for L-interpreter Experimentation. Continued, next page.

```
    |
    | form - F <= \X. one?:X -> X, (mpy:X):(F:(dcr:X)).
    |
tstL = (LBL F
        (LAM X
            (CND (APL (IDE one?) (IDE X))
                 (TST (IDE X)
                      (APL (APL (IDE mpy) (IDE X))
                           (APL (IDE F) (APL (IDE dcr) (IDE X)))))))).

fac:x =: try:<QAPL tstL <QNUM x>>.


    |
    | form - F <= \X. zed?:X -> 1,
    |                one?:X -> 1, (add:(F:(dcr:x))):(dcr:(dcr:X)).
    |
tstN = (LBL F
        (LAM X
        (CND (APL (IDE zed?) (IDE X))
             (TST (NUM 1)
                  (CND (APL (IDE one?) (IDE X))
                       (TST (NUM 1)
                            (APL (APL (IDE add)
                                      (APL (IDE F)
                                           (APL (IDE dcr) (IDE X))))
                                 (APL (IDE F)
                                      (APL (IDE dcr)
                                           (APL (IDE dcr) (IDE X)))))
                       ))))).

fib:x =: try:<QAPL tstN <QNUM x>>.
```

Trial Forms for L-interpreter Experimentation (cont'd). The forms **tstI**, **tstL**, and **tstN** define the *greatest-common-divisor*, *factorial*, and *Fibonacci* functions. The help functions **gcd**, **fac**, and **fib** build applications for repeated testing.

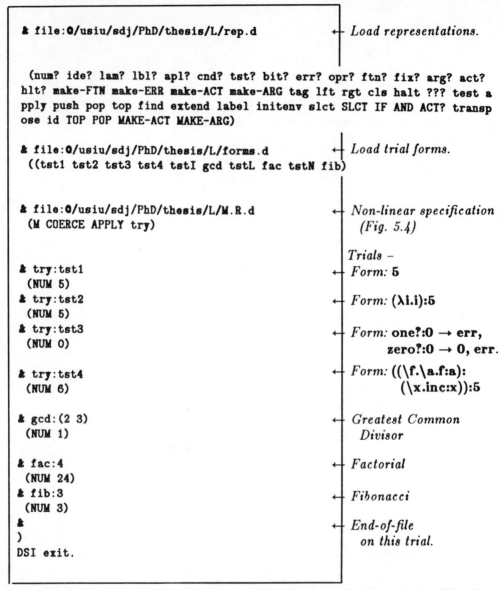

```
& file:0/usiu/sdj/PhD/thesis/L/rep.d          ←| Load representations.


  (num? ide? lam? lbl? apl? cnd? tst? bit? err? opr? ftn? fix? arg? act?
  hlt? make-FTN make-ERR make-ACT make-ARG tag lft rgt cls halt ??? test a
  pply push pop top find extend label initenv slct SLCT IF AND ACT? transp
  ose id TOP POP MAKE-ACT MAKE-ARG)

& file:0/usiu/sdj/PhD/thesis/L/forms.d         ←| Load trial forms.
  ((tst1 tst2 tst3 tst4 tstI gcd tstL fac tstN fib)


& file:0/usiu/sdj/PhD/thesis/L/M.R.d           ←| Non-linear specification
  (M COERCE APPLY try)                              (Fig. 5.4)

                                                 Trials –
& try:tst1                                     ←| Form: 5
  (NUM 5)
& try:tst2                                     ←| Form: (λi.i):5
  (NUM 5)
& try:tst3                                     ←| Form: one?:0 → err,
  (NUM 0)                                             zero?:0 → 0, err.

& try:tst4                                     ←| Form: ((\f.\a.f:a):
  (NUM 6)                                              (\x.inc:x)):5

& gcd:(2 3)                                    ←| Greatest Common
  (NUM 1)                                             Divisor

& fac:4                                        ←| Factorial
  (NUM 24)
& fib:3                                        ←| Fibonacci
  (NUM 3)
&                                              ←| End-of-file
)                                                   on this trial.
DSI exit.
```

Annotated Listing of Trial Runs on the Various Interpreters. The first
trial tests the non-linear specification of the L-interpreter. All trials are on the
expression forms defined on the preceding pages. Trials are continued on the following three pages.

```
& file:0/usiu/sdj/PhD/thesis/L/M.S.d  ←
  (M COERCE RETURN APPLY try)

& try:tst1                            ←
  (NUM 5)
& try:tst2                            ←
  (NUM 5)
& try:tst3                            ←
  (NUM 0)
& try:tst4                            ←
  (NUM 6)
& gcd:(2 3)                           ←
  (NUM 1)
& fac:4                               ←
  (NUM 24)
& fib:3                               ←
  (NUM 3)
&                                     ←
)
DSI exit.

& file:0/usiu/sdj/PhD/thesis/L/M.I1.d ←
  (M tr try)

& try:tst1                            ←
  (NUM 5)
& try:tst2                            ←
  (NUM 5)
& try:tst3                            ←
  (NUM 0)
& try:tst4                            ←
  (NUM 6)
& gcd:(2 3)                           ←
  (NUM 1)
& fac:4                               ←
  (NUM 24)
& fib:3                               ←
  (NUM 3)
&                                     ←
)
DSI exit.
```

(Initialization Deleted)
Stacking specification.
  (Fig. 5.5)

5

$(\lambda i.i){:}5$

one?:0 → err, zero?:0 → 0, err

$((\backslash f.\backslash a.f{:}a){:})\backslash x.inc{:}x))){:}5$

Greatest Common Divisor

Factorial

Fibonacci

End-of-file on this trial.

(Initialization Deleted)
First loop version.
  (Fig. 5.6)

5

$(\lambda i.i){:}5$

one?:0 → err, zero?:0 → 0, err

$((\backslash f.\backslash a.f{:}a){:})\backslash x.inc{:}x))){:}5$

Greatest Common Divisor

Factorial

Fibonacci

End-of-file on this trial.

Trial Runs (cont'd).

```
& file:0/usiu/sdj/PhD/thesis/L/M.I2.d        (Initialization Deleted)
  (M try)                                ←    Refined loop.
                                              (Fig. 5.7)

& try:tst1                               ←    5
  (NUM 5)
& try:tst2                               ←    (λi.i):5
  (NUM 5)
& try:tst3                               ←    one?:0 → err, zero?:0 → 0, err
  (NUM 0)
& try:tst4                               ←    ((\f.\a.f:a):)\x.inc:x)):5
  (NUM 6)
& gcd:(2 3)                              ←    Greatest Common Divisor
  (NUM 1)
& fac:4                                  ←    Factorial
  (NUM 24)
& fib:3                                  ←    Fibonacci
  (NUM 3)
&                                        ←    End-of-file on this trial.
)
DSI exit.

& file:0/usiu/sdj/PhD/thesis/L/M.C1.d         (Initialization Deleted)
& (M monitor try)                        ←    First realization.
                                              (Fig. 5.9)

& try:tst1                               ←    5
  (NUM 5)
& try:tst2                               ←    (λi.i):5
  (NUM 5)
& try:tst3                               ←    one?:0 → err, zero?:0 → 0, err
  (NUM 0)
& try:tst4                               ←    ((\f.\a.f:a):)\x.inc:x)):5
  (NUM 6)
& gcd:(2 3)                              ←    Greatest Common Divisor
  (NUM 1)
& fac:4                                  ←    Factorial
  (NUM 24)
& fib:3                                  ←    Fibonacci
  (NUM 3)
&                                        ←    End-of-file on this trial.
)
DSI exit.
```

Trial Runs (cont'd).

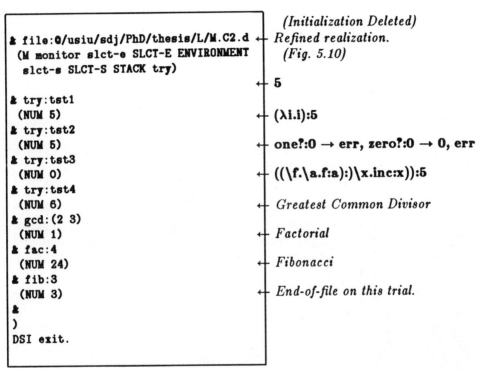

```
& file:0/usiu/sdj/PhD/thesis/L/M.C2.d  ←  (Initialization Deleted)
  (M monitor slct-e SLCT-E ENVIRONMENT    Refined realization.
  slct-s SLCT-S STACK try)                  (Fig. 5.10)

                                       ←  5
& try:tst1
  (NUM 5)                              ←  (λi.i):5
& try:tst2
  (NUM 5)                              ←  one?:0 → err, zero?:0 → 0, err
& try:tst3
  (NUM 0)                              ←  ((\f.\a.f:a):)\x.inc:x)):5
& try:tst4
  (NUM 6)                              ←  Greatest Common Divisor
& gcd:(2 3)
  (NUM 1)                              ←  Factorial
& fac:4
  (NUM 24)                             ←  Fibonacci
& fib:3
  (NUM 3)                              ←  End-of-file on this trial.
&
)
DSI exit.
```

Trial Runs (cont'd). In the realization trials, the help function **try** is redefined to return the content of the VAL register as soon as the *halt* action appears in the EXP register.

```
& file:0/usiu/sdj/PhD/thesis/L/M.R.d          ←  Non-linear specifiation
  (M COERCE APPLY try)                            Load interpreter
& try:(APL (APL (IDE add) (NUM 2)) (NUM 2))    ←  Evaluate 5 : (x ⟸ x)
  (ERR (invalid function))                     ←  Interpretation con-
&                                                 verges to an
)                                                 error message.
DSI exit.



                                                  Circuit realization
& file:0/usiu/sdj/PhD/thesis/L/M.C2.d          ←  Load interpreter

  (M monitor slct-e SLCT-E ENVIRONMENT
   slct-s SLCT-S STACK try)

& try:(APL (NUM 5) (LBL X (IDE X)))            ←  Evaluate 5 : (x ⟸ x)
                                               ←  Tracing CTL, VAL,
(EVL []        (APL (NUM 5) (LBL X (IDE X)))  (HLT)    EXP, and NXT.
 EVL ???       (NUM 5)              ARG (LBL X (IDE X)))
 EVL (NUM 5)   (ARG (LBL X (IDE X))) (HLT)
 EVL ???       (LBL X (IDE X))      (ACT (NUM 5))
 EVL ???       (IDE X)              (ACT (NUM 5))    ←  Evaluator loops
 EVL ???       (FIX (IDE X) beta)   (ACT (NUM 5))
 EVL ???       (IDE X)              (ACT (NUM 5))    ←
 EVL ???       (FIX (IDE X) beta)   (ACT (NUM 5))
 EVL ???       (IDE X)              (ACT (NUM 5))    ←
 EVL ???       (FIX (IDE X) beta)   (ACT (NUM 5))
 EVL ???       (IDE X)              (ACT (NUM 5))    ←
 EVL ???       (FIX (IDE X) beta)   (ACT (NUM 5))
 EVL ???       (IDE X)              (ACT (NUM 5))    ←
 EVL ???       (FIX (IDE X) beta)   (ACT (NUM 5))
 EVL ???       (IDE X)              (ACT (NUM 5))    ←
 EVL ???       (FIX (IDE X) beta)   (ACT (NUM 5))
 EVL ???       (IDE X)              (ACT (NUM 5)) ↑C ←  Daisy interrupted.
```

Demonstration that the Realization is Partial. The literal **beta** is Daisy's symbol for a (non-printable) function closure. Here the object is the environment field of an L-function closure. The realization diverges because the derived interpreter is applicative order.

```
& file:Q/usiu/sdj/PhD/thesis/L/M.C2.d                    ← Load
 (M monitor slct-e SLCT-E ENVIRONMENT
  slct-s SLCT-S STACK try)
& tst                                                    ← Test
 (APL (LBL F                                               form
        (LAM X
          (CND (APL (IDE zed?) (IDE X))
               (TST (IDE X)
                    (APL (IDE F) (APL (IDE dcr) (IDE X)))))))
     (NUM 2))
& try:tst                                                ← Try it.

(EVL []           (APL (LBL F -*-) (NUM 2))  (HLT)
 EVL ???          (LBL F (LAM X -*-))         (ARG (NUM 2))
 EVL ???          (LAM X -*-)                 (ARG (NUM 2))
 EVL (FTN X -*-)  (ARG (NUM 2))               (HLT)
 EVL ???          (NUM 2)                     (ACT (FTN X -*- beta))
 EVL (NUM 2)      (ACT (FTN X -*- beta))      (HLT)
 APL (NUM 2)      (FTN X -*- beta)            (HLT)
 EVL ???          (CND (APL -*-) -*-)         (HLT)
 EVL ???          (APL (IDE zed?) (IDE X))    (TST -*-)
 EVL ???          (IDE zed?)                  (ARG (IDE X))
 EVL ???          (OPR -*-)                   (ARG (IDE X))    ← SPN?
 EVL (OPR -*-)    (ARG (IDE X))               (TST -*-)
 EVL ???          (IDE X)                     (ACT (OPR -*-))
 EVL ???          (NUM 2)                     (ACT (OPR -*-))  ← SPN?
 EVL (NUM 2)      (ACT (OPR -*-))             (TST -*-)
 APL (NUM 2)      (OPR -*-)                   (TST -*-)
 EVL (BIT [])     (TST (IDE X) (APL -*-))     (HLT)
 EVL ???          (APL (IDE F) (APL -*-))     (HLT)
 EVL ???          (IDE F)                     (ARG (APL -*-))
 EVL ???          (FIX (LAM X -*- beta)       (ARG (APL -*-))  ← SPN!
 EVL ???          (LAM X -*-)                 (ARG (APL -*-))
 EVL (FTN X -*-)  (ARG (APL -*-))             (HLT)
 EVL ???          (APL (IDE dcr) (IDE X))     (ACT (FTN X -*- beta))
 EVL ???          (IDE dcr)                   (ARG (IDE X))
 EVL ???          (OPR -*-)                   (ARG (IDE X))    ← SPN?
 EVL (OPR -*-)    (ARG (IDE X))               (ACT (FTN X -*- beta))  (cont'd)
```

Experiment with the Realization.

The form $(F \Leftarrow (\lambda x.\ \text{zero?} : x \to x, F : (\text{dcr} : x))) : 2$ is evaluated to show cycles wasted in testing for expression closures. Signals CTL, VAL, EXP, and NXT are traced. The symbol -*- appears in place of output text that was manually deleted. Useless tests are indicated by the annotation "← SPN?".

188

(cont'd)

EVL ???	(IDE X)	(ACT (OPR -*-))
EVL ???	(NUM 2)	(ACT (OPR -*-))
EVL (NUM 2)	(ACT (OPR -*-))	(ACT (FTN X -*- beta)) ← *SPN?*
APL (NUM 2)	(OPR -*-)	(ACT (FTN X -*- beta))
EVL (NUM 1)	(ACT (FTN X -*- beta))	(HLT)
APL (NUM 1)	(FTN X -*- beta)	(HLT)
EVL ???	(CND (APL -*-) (TST -*-))	(HLT)
EVL ???	(APL (IDE zed?) (IDE X))	(TST -*-)
EVL ???	(IDE zed?)	(ARG (IDE X))
EVL ???	(OPR -*-)	(ARG (IDE X)) ← *SPN?*
EVL (OPR -*-)	(ARG (IDE X))	(TST -*-)
EVL ???	(IDE X)	(ACT (OPR -*-))
EVL ???	(NUM 1)	(ACT (OPR -*-)) ← *SPN?*
EVL (NUM 1)	(ACT (OPR -*-))	(TST -*-)
APL (NUM 1)	(OPR -*-)	(TST -*-)
EVL (BIT [])	(TST (IDE X) (APL -*-))	(HLT)
EVL ???	(APL (IDE F) (APL -*-))	(HLT)
EVL ???	(IDE F)	(ARG (APL -*-))
EVL ???	(FIX (LAM X -*-) beta)	(ARG (APL -*-)) ← *SPN!*
EVL ???	(LAM X -*-)	(ARG (APL -*-))
EVL (FTN X -*-)	(ARG (APL -*-))	(HLT)
EVL ???	(APL (IDE dcr) (IDE X))	(ACT (FTN X -*- beta))
EVL ???	(IDE dcr)	(ARG (IDE X))
EVL ???	(OPR -*-)	(ARG (IDE X)) ← *SPN?*
EVL (OPR -*-)	(ARG (IDE X))	(ACT (FTN X -*- beta))
EVL ???	(IDE X)	(ACT (OPR -*-))
EVL ???	(NUM 1)	(ACT (OPR -*-)) ← *SPN?*
EVL (NUM 1)	(ACT (OPR -*-))	(ACT (FTN X -*- beta))
APL (NUM 1)	(OPR -*-)	(ACT (FTN X -*- beta))
EVL (NUM 0)	(ACT (FTN X -*- beta))	(HLT)
APL (NUM 0)	(FTN X -*- beta)	(HLT)
EVL ???	(CND (APL -*-)) (TST -*-)	(HLT)
EVL ???	(APL (IDE zed?) (IDE X))	(TST -*-)
EVL ???	(IDE zed?)	(ARG (IDE X))
EVL ???	(OPR -*-)	(ARG (IDE X)) ← *SPN?*

(cont'd)

Experiment with the Realization (cont'd).

```
EVL (OPR -*-)      (ARG (IDE X))          (TST -*-)
EVL ???            (IDE X)                (ACT (OPR -*-))
EVL ???            (NUM 0)                (ACT (OPR -*-))      ← SPN?
EVL (NUM 0)        (ACT (OPR -*-))        (TST -*-)
APL (NUM 0)        (OPR -*-)              (TST -*-)
EVL (BIT true)     (TST (IDE X) -*-)      (HLT)
EVL ???            (IDE X)                (HLT)
EVL ???            (NUM 0)                (HLT)                ← SPN?
EVL (NUM 0)        (HLT)                  #                   ← Halt.
)
#
) DSI exit.
```

(cont'd)

Experiment with the Realization (cont'd).

C. Proofs

COROLLARY 2.3-2. *Let FIB and G be defined by*

$$FIB(x) \Leftarrow (x \leq 1) \rightarrow 1, \; FIB(x-2) + FIB(x-1)$$

$$G(x, y, z) \Leftarrow (x = 0) \rightarrow y, \; G(x-1, z, y+z).$$

Then for all $a \geq 0$, $FIB(a) = G(a, 1, 1)$.

PROOF: Using induction hypothesis *"If $a \leq k+1$ then $FIB(a) = G(a, 1, 1)$".*
Base Step.

$$
\begin{array}{l|l}
FIB(0) = 1 = G(0, 1, 1) & \Delta FIB, \; \Delta G \\
FIB(1) = 1 = G(0, 1, 2) = G(1, 1, 1) & \Delta FIB, \; \triangledown, \; \Delta G
\end{array}
$$

Induction. Suppose $0 \leq a \leq k+1$ implies $FIB(a) = G(a, 1, 1)$. Then

$$
\begin{array}{l|l}
FIB(k+2) = FIB(k) + FIB(k+1) & \Delta FIB \\
 = G(k, 1, 1) + G(k+1, 1, 1) & I.H., \; \textit{used twice} \\
 = G(k+2, 1, 1) & \textit{Proposition 2.3-1}
\end{array}
$$

$\square$

COROLLARY 2.3-5. *Let FAC and G be defined by*

$$FAC(x) \Leftarrow (x = 0) \rightarrow 1, \; x * FAC(x-1).$$

$$G(x, y) \Leftarrow (x = 0) \rightarrow y, \; G(x-1, x*y).$$

Then for all $a \geq 0$, $FAC(a) = G(a, 1)$.

PROOF: by structural induction on *Int.*

Base. $FAC(0) = 1 = G(0, 1)$

Induction. Suppose $FAC(k) = G(k, 1)$. Multiplication is commutative and associative, and the equation for G is an instance of the recursion scheme of Proposition 2.3-3. Hence

$$\begin{aligned}
FAC(k+1) &= (k+1) * FAC(k) & &\triangle FAC \text{ and } (k+1) \neq 0. \\
&= (k+1) * G(k, 1) & &I.H. \\
&= G(k, (k+1) * 1) & &\text{Proposition 2.3-3} \\
&= G(k+1, 1) & &\triangledown G \text{ and } (k+1) \neq 0
\end{aligned}$$

$\square$

PROPOSITION 2.4-3. *Consider the linear recursion scheme:*

$$F(x) \Leftarrow p(x) \rightarrow f(x), h(F(g(x))$$

and the iterative system

$$G(x, y) \Leftarrow p(x) \rightarrow H(y, f(x)), G(g(x), y).$$
$$H(x, y) \Leftarrow p(x) \rightarrow x, H(g(x), h(y)).$$

For all a, $F(a) = G(a, a)$.

PROOF:

CLAIM A: *If $p(a)$ is false then for all a and b, $H(a, b) = h(H(g(a)), b)$.*

PROOF: By subgoal induction on H. Since $p(a)$ is false, $H(a, b) = H(g(a), h(b))$. Now if $p(g(a))$ is true then

$$H(g(a), h(b)) = h(b) = h(H(g(a), b))$$

Otherwise, by induction

$$H(g(a), h(b)) = h(H(g(g(a)), h(b))) = h(H(g(a), b))$$

This proves Claim A.

CLAIM B. *If $p(b)$ is false then $G(a, b) = h(G(a, g(b)))$.*

PROOF: By subgoal induction on G. If $p(a)$ is true then $G(a, b) = H(b, f(a))$, which by Claim A equals $h(H(g(b), f(a)))$, since $p(b)$ is assumed false. Under the premise that $p(a)$ is true, this folds to $h(G(a, g(b)))$.

If $p(a)$ is false then $G(a, b) = G(g(a), b) = h(G(g(a), g(b)))$ by induction. However, if $p(a)$ is false, this also folds to $h(G(a, g(b)))$. This proves Claim B.

PROOF of the PROPOSITION: To show that for all a, $G(a, a) = F(a)$, we proceed by subgoal induction on G. If $p(a)$ is true then

$$G(a, a) = H(a, f(a)) = f(a) = F(a).$$

Otherwise,

$$
\begin{array}{ll}
G(a, a) = G(g(a), a) & \Delta G \\
\qquad = h(G(g(a), g(a))) & \text{Claim B} \\
\qquad = h(F(g(a))) & \text{I.H.} \\
\qquad = F(a) & \nabla F
\end{array}
$$

$\square$

THEOREM 2.4-5. *Let F be defined by*

$$F(x) \Leftarrow p(x) \rightarrow f(x), h(x, F(g(x))).$$

and consider the specification.

$$
\begin{aligned}
G(u, v, x, y, z) \Leftarrow p(x) \rightarrow \ & L(u, \phi, u, \phi, fx), \\
& G(u, \phi, gx, \phi, \phi).
\end{aligned}
$$

$$L(u, v, x, y, z) \Leftarrow p(x) \rightarrow z, M(u, gx, gx, u, z).$$

$$
\begin{aligned}
M(u, v, x, y, z) \Leftarrow p(x) \rightarrow \ & L(u, \phi, v, \phi, h(y, z)), \\
& M(u, v, gx, gy, z).
\end{aligned}
$$

For all a, $F(a) = G(a, \phi, a, \phi, \phi)$.

DISCUSSION: Let g^n denote the n-fold composition of g with itself. Observe that if F converges on a, the result is of the form

$$h(a, h(ga, ..., h(g^{(n-1)}a, fg^n a)...))$$

(Some parentheses have been suppressed.) We adopt the following strategy for computing this term iteratively:

(1) Compute $fg^n a$ and call it z.

(2) For $i = n-1, ..., 0$, compute $y = g^i a$ and set z to $h(y, z)$.

The problem is to perform the loop in step (2) without the benefit of a counter. This is done by noting that n is precisely the number of times g must be applied to x in order to make p true. The strategy is implemented by introducing a "trailer" identifier that lags behind the computation of $g^n x$ by i steps, so that when $g^n x$ becomes true, the trailer contains $g^{(n-i)}x$. This value makes it possible to reconstruct the ith outer call. The solution scheme uses five identifiers

u – the initial value of the argument

v – a restart value for the next pass through step (2)

x – the value tested by p .

y – the trailer identifier

z – a value accumulator

As the statement of the theorem asserts, an iterative equation for F is

$$G(u,\ v,\ x,\ y,\ z) \Leftarrow p(x) \rightarrow L(u,\ \phi,\ u,\ \phi,\ fx),$$
$$G(u,\ \phi,\ gx,\ \phi,\ \phi).$$
$$L(u,\ v,\ x,\ y,\ z) \Leftarrow p(x) \rightarrow z,\ M(u,\ gx,\ gx,\ u,\ z).$$
$$M(u,\ v,\ x,\ y,\ z) \Leftarrow p(x) \rightarrow L(u,\ \phi,\ v,\ \phi,\ h(y,\ z)),$$
$$M(u,\ v,\ gx,\ gy,\ z).$$

G computes the inner term $fg^n x$, then resets x to its initial value for the first pass through the loop. L advances x by one step, saves that value for the next pass through the loop, and sets the trailer to x's initial value. M computes $g^{(n-i)}x$ by advancing x and y in tandem.

CLAIM A: If $p(b)$ is false then $M(a,\ b,\ c,\ d,\ e) = h(a,\ M(ga,\ gb,\ c,\ d,\ e))$.

PROOF: By subgoal induction on M, depending on the value of $p(c)$ and $p(gb)$. If $p(c)$ is false, then

$$l.h.s. = M(a,\ b,\ gc,\ gd,\ e) \qquad \qquad \Delta M;\ \neg p(c)$$
$$= h(a,\ M(ga,\ gb,\ gc,\ gd,\ e)) \qquad I.H.;\ \neg p(b)$$
$$= r.h.s. \qquad \qquad \qquad \qquad \qquad \nabla M;\ \neg p(c)$$

Otherwise, if $p(c)$ is true, then

$$l.h.s. = L(a,\ \phi,\ b,\ \phi,\ h(d,\ e)) \qquad \Delta M;\ p(c)$$
$$= M(a,\ gb,\ gb,\ a,\ h(d,\ e)) \qquad \Delta L;\ \neg p(b)$$

and

$$r.h.s = h(a,\ L(ga,\ \phi,\ gb,\ \phi,\ h(d,\ e))) \qquad \Delta M;\ p(c)$$

Now if $p(gb)$ is true, both sides reduce to $h(a,\ h(d,\ e))$. If not, then

$$l.h.s. = h(a,\ M(ga,\ ggb,\ gb,\ a,\ h(d,\ e))) \qquad I.H.;\ \neg p(gb)$$
$$= h(a,\ M(ga,\ ggb,\ ggb,\ ga,\ h(d,\ e))) \qquad \Delta M;\ \neg p(gb)$$
$$= h(a,\ L(ga,\ \phi,\ gb,\ \phi,\ h(d,\ e))) \qquad \nabla L;\ \neg p(gb)$$
$$= r.h.s.$$

This proves Claim A.

CLAIM B: *If $p(a)$ is false, then $G(a,\ \phi,\ c,\ \phi,\ \phi) = h(a,\ G(ga,\ \phi,\ c,\ \phi,\ \phi))$.*

PROOF: By subgoal induction on G; the case depends on the value of $p(c)$.

CASE 1 *($p(c)$ is false).*

$$l.h.s. = G(a,\ \phi,\ gc,\ \phi,\ \phi) \qquad \Delta G$$
$$= h(a,\ G(ga,\ \phi,\ gc,\ \phi,\ \phi)) \qquad I.H.$$
$$= r.h.s. \qquad \qquad \qquad \qquad \nabla G$$

CASE 2 *($p(c)$ is true).*

$$l.h.s = L(a,\ \phi,\ a,\ \phi,\ fc) \qquad \Delta G$$
$$= M(a,\ ga,\ ga,\ a,\ fc) \qquad \Delta L;\ \neg p(a)$$

And on the right,

$$r.h.s = h(a, L(ga, \phi, ga, \phi, fc)) \qquad | \Delta G$$

Now if $p(ga)$ is true, both sides reduce to $h(a, fc)$; so suppose $p(ga)$ is false. Then

$$
\begin{aligned}
l.h.s &= M(a, ga, gga, ga, fc) & & \Delta M; \neg p(ga) \\
&= h(a, M(ga, gga, gga, ga, fc)) & & Claim\ A \\
&= h(a, L(ga, \phi, ga, \phi, fc)) & & \nabla L; \neg p(ga) \\
&= r.h.s. & & \nabla G
\end{aligned}
$$

This proves Claim B.

PROOF of the THEOREM: To show that for all a, $G(a, \phi, a, \phi, \phi) = F(a)$, we proceed by subgoal induction on F. In case $p(a)$ is true, both sides reduce to $f(a)$. If $p(a)$ is false, then

$$
\begin{aligned}
G(a, \phi, a, \phi, \phi) &= G(a, \phi, ga, \phi, \phi) & & \Delta G; \neg p(a) \\
&= h(a, G(ga, \phi, ga, \phi, \phi)) & & Claim\ B \\
&= h(a, G(ga)) & & \nabla G; \neg p(a) \\
&= h(a, F(ga)) & & I.H. \\
&= F(a) & & \nabla F; \neg p(a)
\end{aligned}
$$

$\square$

EXAMPLE 2.4-8

Let us introduce notation to abbreviate the stack operations. For values a and b, and stack σ, let the expression "$[a\ b\ !\ \sigma]$" be an abbreviation for "$push(a, push(b, \sigma))$". Take the formal parameter "$[u\ v\ !\ \sigma']$" to mean that the identifiers u and v name the top two elements of the current stack and σ' names the current stack, understood to be called σ, with u and v removed.

Applications of 1-place operations are abbreviated by suppressing parentheses around the argument. For example, we shall write $g_0(x)$ simply as $g_0 x$. The initial specification is

$$S_0 \quad \boxed{F(x) \Leftarrow p(x) \to c, \, h(\, F(g_0\, x), \, F(g_1\, x)\,)\,.}$$

The first step of the transformation introduces the stack and a return function R.

$$S_1 \quad \boxed{\begin{array}{l} F(x, \sigma) \Leftarrow R((p(x) \to c, \, h(\, F(g_0\, x), \, F(g_1\, x))\,)\,), \, \sigma)\,. \\[1mm] R(x, \sigma) \Leftarrow empty?(\sigma) \to v, \, \cdots \,. \end{array}}$$

Use Rule 2 to distribute R through the conditional.

$$S_2 \quad \boxed{\begin{array}{l} F(x, \sigma) \Leftarrow p(x) \to R(c, \sigma), \, R(\, h(\, F(g_0\, x), \, F(g_1\, x)\,), \, \sigma)\,. \\[1mm] R(v, \sigma) \Leftarrow empty?(u) \to v, \, \cdots \,. \end{array}}$$

Define $e' = h(y_1, F(y_2\,))$; $r = F(g_0\, x)$; and $t = g_1\, x$. Allocate an action value, $a = 0$. By Rule 3, transform S_2 to

$$S_3 \quad \boxed{\begin{array}{l} F(x, \sigma) \Leftarrow p(x) \to R(c, \sigma), \, R(\, F(g_0\, x), \, [0\, g_1\, x\, !\, \sigma]\,)\,. \\[1mm] R(v, \, [w\, z\, !\, \sigma'\,]) \Leftarrow empty?(\sigma) \to v, \\[1mm] \qquad\qquad\qquad at?(w, \, 0) \to R(\, h(v, \, F(z)), \, \sigma'\,), \, \cdots \,. \end{array}}$$

By Rule 1 we can get rid of the second call to R in the equation for F.

$$S_4 \quad \boxed{\begin{array}{l} F(x, \sigma) \Leftarrow p(x) \to R(c, \sigma), \, F(\, g_0\, x, \, [0\, g_1\, x\, !\, \sigma]\,)\,. \\[1mm] R(v, \, [w\, z\, !\, \sigma'\,]) \Leftarrow empty?(\sigma) \to v, \\[1mm] \qquad\qquad\qquad at?(w, \, 0) \to R(\, h(v, \, F(z)), \, \sigma'\,), \, \cdots \,. \end{array}}$$

Let $e' = h(y_1, \, y_2\,)$; $r = F(z)$; and $t = v$. Make the final transformation according to Rule 3, with new action value $a = 1$.

$$S_5 \quad \boxed{\begin{aligned} & F(x, \sigma) \Leftarrow p(x) \to R(c, \sigma),\ F(\,g_0\,x,\ [0\ g_1\ x\ !\ \sigma]\,). \\ & R(v, [w\ z\ !\ \sigma']) \Leftarrow \text{empty?}(\sigma) \to v, \\ & \qquad\qquad\qquad at?(w,\ 0) \to F(z,\ [1,\ v\ !\ \sigma']) \\ & \qquad\qquad\qquad at?(w,\ 1) \to R(\,h(v,\ z),\ \sigma'\,). \end{aligned}}$$

□

PROPOSITION 4.2-1 *For all environments ρ, and all expressions e and e',*

$$D[\![\ (\lambda\,[\,h\,!\,t\,]\,.\,h\,):\,<\,e\,!\,e'>\ ]\!]\rho = D[\![\ e\ ]\!]\rho$$

and

$$D[\![\ (\lambda\,[\,h\,!\,t\,]\,.\,t\,):\,<\,e\,!\,e'>\ ]\!]\rho = D[\![\ e'\ ]\!]\rho$$

PROOF: Both assertions have similar proofs, differing only in the last few steps. Only the second proof is given here. Consult Figure 4.4 for the relevant function definitions.

$$
\begin{aligned}
& D[\![\ (\lambda\,[\,h\,!\,t\,]\,.\,t\,):\,<e\,!\,e'\,>\ ]\!]\rho & \\
& = \textit{d-apply}\ (D[\![\ \lambda\,[\,h\,!\,t\,]\,.\,t\ ]\!]\rho)\ (D[\![\ <e\,!\,e'\,>\ ]\!]\rho) & \Delta D \\
& = \textit{d-apply}\ (\lambda\,v\,.\,D[\![\ t\ ]\!]\rho[\ v\,/\,[\text{h!t}]\ ])\,\langle D[\![\ e\ ]\!]\rho\,,\,D[\![\ e'\ ]\!]\rho\rangle & \Delta D,\ \textit{twice} \\
& = (\lambda\,v\,.\,D[\![\ t\ ]\!]\rho[\ v\,/\,[\text{h!t}]\ ])\,\langle D[\![\ e\ ]\!]\rho\,,\,D[\![\ e'\ ]\!]\rho\rangle & \Delta\,\textit{d-apply} \\
& = D[\![\ t\ ]\!]\rho[\ \langle D[\![\ e\ ]\!]\rho\,,\,D[\![\ e'\ ]\!]\rho\rangle\,/\,[\text{h!t}]\ ] & \textit{substitution} \\
& = \rho[\ \langle D[\![\ e\ ]\!]\rho\,,\,D[\![\ e'\ ]\!]\rho\rangle\,/\,[\text{h!t}]\ ](t) & \Delta D \\
& = \rho[\ D[\![\ e'\ ]\!]\rho\,/\,t\ ][\,D[\![\ e'\ ]\!]\rho\,/\,h\ ](t) & \Delta\rho[\,v\,/\,i\,] \\
& = \rho[\ D[\![\ e'\ ]\!]\rho\,/\,t\ ](t) & \Delta\rho[\,v\,/\,i\,] \\
& = D[\![\ e'\ ]\!]\rho & \Delta\rho[\,v\,/\,i\,]
\end{aligned}
$$

□

PROPOSITION 5.3-1. *For* $\alpha : Env \to Val$,

$$fix\ (\ \lambda\epsilon.\alpha\rho[\ \epsilon\ /\ i\ ]\ ) = \alpha\ (fix\ (\ \lambda\rho'.\rho[\ \alpha\rho'\ /\ i\ ]\ ))$$

PROOF: Let $\rho_0 = fix\ \lambda\rho'.\rho[\ \alpha\rho'\ /\ i\ ]$, and define $v_0 = \alpha\rho_0$. Since ρ_0 is a fixed point,

$$\rho_0 = \rho[\ \alpha\rho_0\ /\ i\ ] = \rho[\ v_0\ /\ i\ ]. \tag{*}$$

Hence,

$$v_0 \overset{def}{=} \alpha\rho_0 = \alpha\rho[\ v_0\ /\ i\ ]$$

and so v_0 is a fixed point of $(\ \lambda\epsilon\ .\ \alpha\ \rho[\ \epsilon\ /\ i\ ]\ )$. Let v_1 be any other fixed point, and define $\rho_1 = \rho[\ v_1\ /\ i\ ]$. Then since v_1 is a fixed point, consider

$$\rho_1 = \rho[\ ^{\alpha\rho[\ v_1\ /\ i\ ]}\ /\ i\ ] = \rho[\ \alpha\rho_1\ /\ i\ ]$$

Thus, ρ_1 is a fixed point of $\lambda\rho'.\ \rho[\ \alpha\rho'\ /\ i\ ]$; hence $\rho_0 \sqsubseteq \rho_1$. By (*) we have

$$\rho[\ v_0\ /\ i\ ] = \rho_0 \sqsubseteq \rho_1 \overset{def}{=} \rho[\ v_1\ /\ i\ ]$$

Since ρ_0 and ρ_1 only differ at i, it must be that

$$\rho[\ v_0\ /\ i\ ](i) = v_0 \sqsubseteq v_1 = \rho[\ v_1\ /\ i\ ](i).$$

That is, v_0 is the minimal fixed point. Therefore,

$$fix(\ \lambda\epsilon.\alpha\rho[\ \epsilon\ /\ i\ ]\ ) = v_0 \overset{def}{=} \alpha\rho_0 \overset{def}{=} \alpha\ (fix\ (\ \lambda\rho'.\rho[\ \alpha\rho'\ /\ i\ ]\ )).$$

$\square$

D. Symbols

a, b, c – constant symbol, 17

x, y, z – identifier, 17

p, q – predicate symbol or propositional expression, 17

r, s, t – term or expression, 17

e – expression, 17

F, G, H – function variable symbol, 17

$p \rightarrow r, s$ – conditional expression, 18

$F(x) \Longleftarrow r.$ – recursive function definition, 19

ϕ – indeterminate, or "floating" value, 16

$!$ – register initializer, 52

$\bullet$ – value token in a schematic, 54

$\boxed{f}$ – component counterpart to an operation symbol, 52

$\boxed{!}$ – register, 54

Ξ – identity component, 129

$[\![\, ... \,]\!]$ – syntactic quotation, 17

$A + B$ – separated sum domain, 42

$A \times B$ – product domain, 42

$A \rightarrow B$ – continuous function domain, 42

$f{:}A{\rightarrow}B$ – $f \in A{\rightarrow}B$, 43

$\langle ... \rangle$ – pairing operation, 43

$*{\downarrow}0, *{\downarrow}1, ...$ – projection for domain pairs, 43

$*isD, *asA, *inD$ – inspection, restriction, and injection for domain sums, 43

E – domain element designator, 41

$\sqsubseteq$ – approximation ordering, 41

$\perp$ – minimal approximation, 41

$\lambda x.e$ – abstraction of e by x, 42

$\lambda[u\ v]\,.\,e$ – $\lambda x.e(x\!\downarrow\!0)(x\!\downarrow\!1)$, 44

$\equiv$ – strongly equivalent, 73

Sig_D – the domain of signals over D, 63

d^{∞} – constant signal, 64

ΔF – "by unfolding F's definition", 22

∇F – "by folding F's definition", 22

$I.H.$ – "by induction hypothesis", 27

Φ_F – verification condition for F, 25

x^0 – initial value of signal X, 56

$X^{@n}$ – behavior of signal X at time n, 53

K^c – constant operation, 31

π – projection operation, 31

$f{\circ}g$ – serial combination (*i.e.* composition) of operations, 31

$<...>$ – parallel combination of operations, 32

$e\begin{bmatrix} t_1,\,...,\,t_n \\ x_1,\,...,\,x_n \end{bmatrix}$ – a substitution, 21

τ – translator from applicative terms to combined operations, 32

$\mathbb{T},\ \mathbb{D},\ \mathbb{L},$ – a valuation function, 46, 70–75, 104

$\mathbb{R}$ – L compiler, 106

Rep_V – a representation for V, 105

$\overset{\triangledown}{exp},\ ^{\triangledown}[...]$ – abstract value of, 105

$[tag\ ...]$ – represented value, 105

$\hat{u}$ – an instance of the value u, 131

Index

A

Abstract component, 13, 89, 93–100
Accumulator, 9, 37
Action, 112
— value, 38
Actual expression, 127
Admit selection, 37
Agent, 93
ALGOL 60, 11, 68
Application in DAISY, 73
applicative
— language, 67
— premise, 1
— style, 1
apply, 44, 49
APPLY component, 95
Approximation ordering, 41
Axioms E, F, G, N, S, 127

B

Backus-Naur notation, 46
Balanced form, 30, 34
behavior, 64
Behavior, 3, 53, 64
— , domain model of, 63–65
— equivalence, 3, 53, 64
— , instantaneous, 51–52
— , relational model of, 144
Bidirectionality, 144
Bool, 41
Branched conditional form, 25, 29

C

Call-by-name, 68
Call-by-value, 68
carrier, 16
C_{FIB}, 98
CHESS, 149
Circuit description, 53
Circuit emulation (in DAISY), 81–85
Circuit F, 132
Circuit G, 133
Circuit H, 136
Circuit refinement, 139, 143
— local, 125
Clock, 51
Combinator, 46
Combinatorial component, 52
Combined operation, 31–33
Communication, 95, 144–147
Compilation, 2
Completely strict, 23
Component, 5, 10, 64
— in DAISY, 81
Computation rule, 68
Conditional expression, 18
Connective storage, 125
Constant, 16
— combinator, 44
— operation, 31
— signal, 64
Continuation, 46
Control token, 34, 142, 146
Convergent term, 23
Creativity, 29, 148

Cuny-Snyder model, 125, 139
curry, 44

D

ID, 73, 76–77, 127
DAISY, 4, 13, 67–79, 142
— kernal language, 71
— standard semantics, 71
— syntax, 69
d-apply, 73, 128
Data recursion, 11, 13, 80, 125
Decomposition of control, 144
Defining equation, 19
Delay rule, 68
— for transformation, 129
Demand driven, 67–68
Device, 82
Dig, 16, 146
Digital asynchrony, 145
Digital circuit design, 1
Digital/synchronous system, 5, 141
Direct interpreter, 101
Distributivity
— of combination over lifting, 57–61
— of the conditional, 30, 48
Domain, 41
— continuous function on 41,
— flat, 41
— function on, 42
— n-ary product, 42
— non-flat, 42
— product, 42
— semantic, 42
— sum, 42
— syntactic, 42
Don't-care, 16
Don't-know, 16

E

Engineering, 2
Environment, 22, 46
— as a data structure, 67

— in DAISY, 72
ENVIRONMENT, 116
Evaluation object, 108
Experiment, 71, 83
Experimentation, 142
— in DAISY, 121
Expression, 18
Expression closure, 68, 108

F

FAC, 20, 22–24, 28, 37, 37, 61, 70, 83–84, 105
— as a fixed point, 45
Feedback, 2, 124
FIB, 20, 37, 61, 70, 84, 97, 105
fix, 44
Fixed point, 44
Flowchart, 6, 141
Flowchartability, 3, 37
Folding, 22
— circuits, 123–124
Formal expression, 126
Formal parameter, 17
FP programming, 9
Function closure, 107
Function variable symbol, 17
Functional, 44

G

GCD, 20, 62, 70, 85, 105, 142, 146
Global identifiers, 19
Gluing, 127–128
GO, 149
Grammatical transformation (begin), 29–30
Graph reduction, 67
Ground term or expression, 19

H

Hierarchical decomposition, 90–98
Higher level components, 90–100

I

Ide, 42
Identification of signals, 62, 132
Identifier, 17, 42
Identity component, 129
Indeterminacy, 147
Indeterminate constant (ϕ), 16
Infix notation (footnote), 19
Injection, 43
Input-output assertion, 25
Inspection, 43
Installation, 130
Instance
   — of a scheme, 20
   — of a value, 131
Instruction, 13, 89, 94, 116, 142
Int, 17, 19
   — domain, 41
Interpretation, 67
Iteration, 9, 141
Iterative
   — form, 15, 28
   — specification, 20, 141

J K

K, 46, 76–77

L

L, 104
L_E, 18
L_R, 18
L_S, 52
L_T, 18
Lambda abstraction, 42
Language driven design, 101
Language *L*, 13, 90, 103–105
   — interpreter, 102, 105–110
Lazy evaluation, 68
Lifting, 6, 54, 141
   — rule, 129
Linear specification, 6
Linear term, expression, specification,
   20
Linearization
   — control, 46
of specifications, 38
LISP, 8, 45, 67, 73, 103 (footnote)
LUCID, 9, 148

M

IM, 107–116
Message, 71
Metalinguistic variable, 17
Minimality, 23
Monadic operations (in DAISY), 81
Multiphase clocking, 147–148
Multiple valued operations, 49
Multiplexor, 30
mux, 30–31

N

IN, 46, 76–77
Nil, 73, 129
Nml, 42
Non-finite data, 80
   — in DAISY, 81
n-place operation, 16

O

Observation, 80
Operation, 16
Output driven computation, 82–83

P

Packaged combination, 89, 92–93, 98,
   142
Packaged component, 13
Padding (of formal arguments), 30
Pairing operation, 43
Parallel combination, 32, 44
Partial correctness, 2, 27, 40
Performance conditions, 148
predicate, 73

predicate, 16
Probe, 73
Process, 10, 65
— autonomous, 145
Projection, 31, 44
PROLOG, 144
Propositional expression, 18

Q R

$I\!R$, 106
Rank, 17
READY, 57
Realization, 2
— language, 12
— of a specification, 57
Recurrent term, 18
Recursion, 44–45
— equation, 15–19
— Recursion scheme, 15– 20
Reduction, 22
Reflexivity, 11, 45
Register, 5, 51–52, 54, 124
— transfer, 54
Rep $_{Env}$, 106
Rep$_{Exp}$, 106
Rep$_{Val}$, 106
Rep$_{Val}$, 110
Representation, 105
Restriction, 43
Rule Δ, 129
Rule G, 128
Rule Λ, 129
Rule M, 130
Rule Φ, 130

S

Schedule, 126, 143
— derivation, 126, 131
Schematic, 5, 54, 124
SCHEME, 67
Scott-Strachey language, 4, 7, 12, 16,
    40–45, 63
Selection, 31

— rule, 130
Self-timing, 145
Semantics, 45
— for terms, 46
— of circuit descriptions, 65
— of DAISY, 71
of L, 103
Sequential control, 3, 15, 28, 31, 37
Serial combination, 31, 44
Serialization, 123–125
Serious, 18, 19
Signal, 10, 51, 63, 142
— equation, 53
— expression, 51–52
Simple loop, 6, 12, 61, 114, 142
Simplification, 22
Single-pulser, 145
Solution
— as a fixed point, 45
— of a specification, 23
Specification, 2, 4, 12, 15, 19
— of control, 46
Stability, 85, 142
Stack, 38
STACK, 94, 116
Standard semantics, 46
State, 9
Storage element (footnote), 124
Stream, 11, 81
Strict, 23
— , completely, 129
strict, 82
Strong equality, 74
Structural induction, 24–25
Structured digital design, 7
Structured programming, 7
Subgoal induction, 25–28
Substitution, 21
Suspension, 11, 68
Synthesis, 2
Synthesis of iterative form, 37

T

T, 46–47
T, 32, 59
Term, 18
— over a set of identifiers, 19
— translator (T), 59
Terminal, 18
time, 52
Transformation, 3, 141
transpose, 64
Trivial, 18, 19
TTL, 15

U

U_I, 28, 33, 36–37, 55–56, 91
U_L, 36
uncurry, 44
Underlying type, 5, 15–16,
Unfolding, 22
Universal scheme, 33–37
Universal type, 15

V

val, 21–24
Valuation, 21–24
VALUE, 57
Value history, 6, 8
Verification condition, 25
Version of a specification, 28

W

Wand-Friedman construction, 38–40,
90, 112

X Y Z